PAPERMAKING
IN BRITAIN
1488–1988

DR RICHARD L. HILLS

PAPERMAKING IN BRITAIN 1488–1988

A Short History

THE ATHLONE PRESS

London and Atlantic Highlands, NJ

First published 1988
by The Athlone Press, 44 Bedford Row, London WC1R 4LY
and 171 First Avenue, Atlantic Highlands, NJ 07716
Copyright © Richard L. Hills 1988

British Library Cataloguing in Publication Data
Hills, R.L. (Richard Leslie), *1936*
    Papermaking in Britain 1488 – 1988.
    1. Great Britain. Paper manufacturing industries, 1488-1988.
    I. Title
    338.4'76762'0941
    ISBN 0-485-11346-5

Library of Congress Cataloguing-in-Publication Data
Hills, Richard Leslie, 1936–
    Papermaking in Britain, 1488–1988.

    Bibliography: p.
    Includes index.
    1. Papermaking—Great Britain—History. I. Title.
TS1095.G7H55   1988   676   88-3322
ISBN 0-485-11346-5

Typeset by Rapid Communications Ltd, London WC1
Typeset in 10½ point Ehrhardt

Jacket illustration: A 'Britannia' watermark sewn onto a laid cover

This book was printed and bound at the University Press, Cambridge. The paper
is Mellotex Smooth Ultra White produced by Tullis Russell and Company Ltd.

The cases are made with boards from the Radcliffe Paper Mill Company,
covered with Linson Buckram from Watson Grange Ltd, a member of the
Tullis Russell Group.

Kraft sleeves supplied by Landor Cartons

CONTENTS

FOREWORD

How easy it is to take familiar things for granted, and what other material is so pervasive in our everyday lives as paper and board in all its many forms – not only the books, magazines and writing papers, but in cars, on our walls and ceilings, in our kitchens and bathrooms, the filters, the tissues, or bank notes – the list is endless.

How easy to overlook the environmental benefits of this infinitely renewable, bio-degradable and recyclable material that in today's computer jargon is so 'user friendly'.

Outside the Industry it is often assumed, quite incorrectly, that because paper has been made in this country for hundreds of years, and in its simplest form can be easily made by hand, it is a process requiring a low level of technology.

The 500th anniversary of paper making in the UK presents an excellent opportunity to survey the development in paper technology over the years and chronicle the milestones in our history. Years that have seen the transition from handmade papers through the traumatic period of the introduction of machines, and the fears that must have been felt by the journeymen who could not possibly envisage the potential explosion in demand, through to the high technology computer controlled environment of today, when we see the Industry re-emerging once again strong, after large sectors were decimated during the 1970s.

Dr Richard Hills, an historian by training and inclination with a deep interest and knowledge of the Industry, has brought together in this book a wealth of information about its technical development over the period, and provides a valuable account of an Industry that contributes so much to our economy and everyday life.

Jeffery Bartlett
Director General
British Paper and Board Industry Federation   London

ACKNOWLEDGEMENTS

My first introduction to the paper industry was twenty years ago when I visited Nash Mills at St. Mary's Cray in Kent to see the exhibits of the National Paper Museum and decide whether we should accept them for the new Museum of Science and Technology in Manchester. The exhibits had been flooded by the river and were in quite a mess but I have never regretted recommending their acceptance because involvement with the paper industry has been one of the most interesting aspects of my life since then.

This book seeks to record a little about the technical achievements of the paper industry and particularly those inventions made here in Britain. I would like to thank all those who have helped with this aspect of my education during the intervening years and I apologise to any whose name is not included here because there have been so many people who have helped in so many ways and who have so kindly shown me around their mills, both in this country and overseas. I have been in a privileged position as President of the International Paper Historians to see many mills on the Continent and also to be helped by colleagues in that Association. I am most grateful to them, and to all those who have allowed me to take up their valuable time, for giving me a broad view of the industry which has been invaluable in the preparation of this book.

The fact that there is a book at all to help celebrate five hundred years of papermaking in Britain we owe to the British Paper and Board Industry Federation and George Mandl. Without their campaigning to obtain the initial backing, the project could not have started. The librarians at the patents section of the Central Library in Manchester and at the University of Manchester Institute of Science and Technology as well as the National Paper Museum Collection have given invaluable assistance. Offers to help with illustrations were received from the Bank of England, Bowaters U.K. Paper Co. Ltd., Comptroller of Her Majesty's Stationery Office, James Cropper, plc., C. Davidson & Sons, DRG Paper & Board, W. Green Son & Waite Ltd., Kimberley Clark Ltd., J. & J. Makin Ltd., Paper Industry Research Association, Purfleet Board Mills, Reed Paper & Board (UK) Ltd., St. Regis Paper Co. (UK) Ltd., Severn Lamb Ltd., Shotton Paper Co. Ltd., Thomas Tait & Sons and Wiggins Teape Ltd. Unfortunately space did not allow us to use all the pictures they offered.

I would like to thank those upon whom I inflicted the typescript for their gentle comments and corrections. These include Edo Loeber, Holland, Dr. Peter Tschudin, Switzerland, as well as colleagues in the Paper Science Department, U.M.I.S.T., and in particular Philip Howarth. Messrs. Green Son and Waite made important contributions with the later sections on watermarking as did Kimberley Clark on tissues and other people on other sections. Some of the latest statistics were provided by the B.P. & B.I.F.

To all these people, and to the many others who have helped and given encouragement in so many ways I am most grateful and I hope that what follows will be worthy of their support in recording a little of the history of a great industry which still is vital to this country and contributes so much to our well-being today.

Richard L. Hills

I

THIS PAPER THYNNE

The Paper-Maker
'The water turns my mill wheel round,
Where rags to paper pulp are ground:
Their snowy leaves on felt I lay,
And squeeze the water well away,
And then I hang my sheets to dry:
All white, and shining like the sky.'
HANS SACKS, about 1550

Why should Henry VII have visited the first paper mill in Britain during May 1498, almost five hundred years ago? We have of course today the example of our royal family going to all sorts of places and industries, but we would hardly expect Henry VII, the founder of Tudor despotism, to be courting popularity among his subjects. Growing royal authority needed records. The laws of the realm had to be written down. Legal proceedings had to be documented. The king wanted to know the state of his finances, what was the taxable potential and what had been raised. Likewise an expanding commerce necessitated keeping accounts, bills of exchange to settle payments, as well as correspondence of all sorts. Some way of preserving all these laws, records and transactions had to be discovered which would not be too bulky and which could be handled easily so that reference could be made to the entries. Paper was one possible substance and perhaps Henry VII visited that infant manufactory because he hoped that it would enable his country to become self-sufficient in a commodity which was assuming greater and greater importance in the legislature and world of commerce.

If we go back to Henry VII's distant predecessor, William the Conqueror, we will discover that he had no alternative but to have his famous Domesday Book of 1086 written on parchment because at that time, no paper was available in England and it was barely known in the rest of Europe. Parchment was made from the skins of animals, in particular sheep. England was fortunate throughout the Middle Ages in having large flocks of sheep which brought prosperity in many ways. Their wool had been exported through the wool trade which had been overtaken and replaced by the cloth trade. At a period when many animals had to be slaughtered at the beginning of winter because there was insufficient fodder, skins for making parchment were probably more readily available than would be the case today.

1

During the fifteenth century, the price of paper had fallen by something like 40 per cent while that of parchment had risen by possibly a similar amount. The quality of parchment had declined at the same time. This movement in favour of paper continued in the sixteenth century for while paper prices rose between 30 to 60 per cent during the first eighty years of that century, that of parchment rose a further 70 per cent.[1] So we find that increasingly paper was replacing parchment during the later part of the Middle Ages.

Paper was important as a wrapping substance as well as a writing surface. Coarse grades of paper and, in particular brown paper, were used as a wrapping material from a very early time but, such is the nature of this use that we have few details of it because most has been thrown away. Our history of papermaking derives principally from white paper because laws, contracts, accounts, etc. have been written on it and preserved in archives. Both from its uses and from the way it has been filed, such paper can be identified and dated with fair accuracy. Generally we find that a country had been importing paper for a considerable period before starting to make paper. England was no exception for, while our white paper industry was not firmly established until the end of the seventeenth century, the oldest surviving piece of paper found in the Public Record Office dates from about 1220. Yet it is not until after 1500 in Tudor times that paper is to be found with any frequency among the records there.[2]

In France, a mention of the Pielle mill at Troyes in 1338 must be considered as confirming the oldest manufactory of paper in that country,[3] although a date of 1326 has been claimed for Ambert.[4] Yet paper had been used in France for at least one hundred years before that. Some paper must have been imported through Marseilles from Syria during the thirteenth century because the Charter of 1237 and the Register of 1249 in the archives at Bouches-du-Rhone are evidence of this.[5] In Italy, a society for manufacturing paper had been formed near Genoa in 1255 but there could have been a paper industry in southern Italy before that. In 1220, Frederick II had forbidden the use of paper for public records in the councils and courts of Naples, Sorrento and Amalfi, while in 1145 the King of Sicily had ordered that deeds written on paper must be transcribed onto parchment because it was feared that paper would not last. The oldest document on paper in the Italian peninsular appears to be one of 1109 in the archives of Palermo in Sicily.

It was the Moors who introduced papermaking into Italy but they had brought it into Europe even earlier through Spain. The traditional date is 1151, by which time there was a flourishing industry at Xativa in Valencia. However, the first paper mill may have been started here as early as 1056 and another one at Toledo in 1086.[6] From this we can see that William the Conqueror had no alternative but to use parchment for the Domesday Book, the more so because the other material commonly used as a substance for writing in the ancient world, papyrus, had disappeared by the tenth century. The reason for this is unknown. It may have been that the papyrus reed had died out in the river Nile because it had to be re-introduced recently to enable modern production to start once more. It may have been that papyrus could not compete with paper being made first in Damascus where it had started in about 794 and a little later in Egypt itself. The traditionally assumed date for the introduction of papermaking to the West is 751 when allegedly the Arabs captured some Chinese papermakers at the battle of Talos near Samarkand and so learnt the art. Yet in 648 the Buddhist monk, Hiuen Tsang had encountered a place about three miles south of the town Taras in Turkish territory populated by some three hundred Chinese families among whom there may have been papermakers.

The art of papermaking had taken a long time to spread beyond the Chinese borders, for the original centre was the Hunan province in the South East of that country. Tradition has it that in 105, T'sai Lun first presented a sheet of paper to the Emperor as a substitute writing surface instead of the silk or other cloth which had been used up to that time. At this point, we must ask 'what is true paper?' The usual definition is that it is a matted material made from cellulose fibres from herbs or trees which have been macerated or beaten to fibrillate them. Then they have to be dispersed in water until they are suspended evenly so that, when the water is drained off, a sheet of paper is left. It is of course possible to make a felted mat of fibres which can look like paper, but one of the hallmarks of true paper must be the beating which breaks down the fibre walls and enables the vital hydrogen bonding to form between the fibres. It is this hydrogen bonding which gives paper most of its cohesion and tearing strength.

Recent research in China using analysis with electron microscopes has shown that what was thought to be some of the oldest paper in the world has never received any beating.[7] In the Baqiao brick factory of the Shanxi province there was discovered a tomb dating from the West Han dynasty (140–87 BC) with, among other objects, three bronze mirrors. Under these mirrors was cloth and what appeared to be paper. Not only is the dating doubtful but the electron microscope has shown that this paper was probably only wadding or even sweepings off the floor. The average fibre length is far too short to have been achieved by any mechanisms available then or much later. Similar investigations into other early papers which have been dated to the pre-Christian era have shown that some of them too are a type of wadding and not true beaten paper. However, those papers made after the time of T'sai Lun do show the effects of beating in their fibre structure to develop hydrogen bonding and compactness of the sheet and are therefore considered to be true paper. Perhaps the legend of T'sai Lun may be the truth after all and perhaps paper was originally developed as a material for writing.

In western Europe the demand for paper was increased by Johann Gutenberg's improvements to printing around 1446 at Mainz in Germany. Up to this time, the letters on each page to be printed had to be carved out of a solid block of wood. Gutenberg cast each letter individually so that, when the printing of that page was completed, the type could be dismantled and used again. By 1452–6 he was able to print a Vulgate Bible. In 1475, William Caxton returned to England determined to establish his own printing press. Caxton had been apprenticed to the Mercers' Company in 1438 to one Robert Large.[8] In 1463, he was appointed Governor of the Company of Merchant Adventurers and spent many years in the Low Countries involved in its affairs there. It was while he was on the Continent that he learnt the art of printing. Members of the Mercers' Company held control over the Merchant Adventurers whose constitution at this period was extraordinarily fluid. While the Merchant Adventurers negotiated as a body with the authorities on matters of commercial policy, chartered ships, arranged convoys to the marts and times of sailing, its members traded individually and remained closely connected with the craft to which they happened to be members. The Governor of the Merchant Adventurers, virtually always a London Mercer, moved constantly from England to the Netherlands and back, and held court at the Mercers' Hall in London. While the Mercers' chief trade was in fine cloth, they dealt in many other commodities, such as sugar and spices, wood, wine, oil, tar, in fact in almost anything that offered a profit. In their records, we find one Robert Tate, possibly John Tate's son, importing one hundred bundles of brown paper in 1497.[9]

Caxton established his printing press at Westminster Abbey where he might have been able to avoid some of the restrictions imposed by the guilds in the City of London. The more probable reason was that Caxton saw a better market for the books he wished to print among the courtiers and noblemen attached to the royal household than among the merchants of the City. What is certain is that by the end of 1476, he had settled all the formalities of finding premises in Westminster and had printed an 'indulgence' which has the date 13 December 1476 added by hand on the only surviving copy. He produced thirty books in the first three years and continued to operate his press until his death in 1491. Then his business was taken over by Wynken de Worde who later moved to Fleet Street in the City of London and became a member of the Stationers' Guild. By 1403, the Stationers had formed their own trade guild. This was incorporated in 1557 as the Stationers' Company and given control of all printing in England.

Gutenberg printed his Bible partly on parchment and partly on paper. It has been estimated that each parchment copy, consisting of more than 641 leaves, needed the skins of more than 300 sheep. In the event, he printed only thirty copies on parchment and 180 copies on paper. By the time of Caxton, an edition of a printed book might run to a thousand copies and so it was evident that paper was the only possible material if the printing industry were to expand. Possibly John Tate had expected a quicker growth in publishing than in fact occurred for it was to be some years before another press was established in England and the first book printed in Edinburgh dates from 1508. However, there was a continuous demand for paper of high quality throughout these years for printing as well as for writing. The first paper that was used in England came from Italy and France from the early fourteenth century until the end of the sixteenth century.[10] The region around Troyes in France formed an important source of supply to the Netherlands and through them to the

English market.[11] Caxton procured his paper from the Low Countries and in it has been found watermarks of a bull's head and a unicorn.[12]

Although people were not to realise it at the time, the accession of Henry VII in 1485 ushered in a time of political stability after the turmoil caused by the Wars of the Roses, allowing trade inside the country to begin to prosper again without fear of disruption. This was essential for papermaking because paper in Europe was made from rags which had first to be collected, then sorted into types of fibre and colour, before being taken to the papermill and such activities needed a peaceful state in the land. At this period, hemp ropes and linen cloth were the principle sources of fibres for papermaking and, to make white paper, only white fibres could be used because there were no bleaches to remove dyes or discolourations. Old sails of ships were one source of supply. Another was the worn-out garments from people. Papermills were therefore established close to large towns where it was easier to collect rags being thrown away by the inhabitants.

Nothing is known about John Tate's reasons for launching out into the paper industry. It may be significant that he also was a member of the Mercers' Company. His father was John Tate, another Mercer, who was Mayor of London in 1473 and died in 1478 or early 1479. It is difficult to identify positively either of these John Tates and the position is further confused because they had a relative, Sir John Tate, who was Mayor in 1496 and died in 1514. He had a son John as well.[13] Our John Tate also had worked for the Merchant Adventurers on the Continent and so was well placed to be involved in trade and commerce. On his return from the Low Countries, he may well have thought that he saw a profitable opening for selling paper to the fledgling printing industry.

Precisely when John Tate began to build his papermill is unknown. J.B. Powell gave a date of 1476[14] and claimed to have found Tate's watermark in one of the Paston letters dated 1485, but no evidence has been found to confirm either of these dates.[15] A rare, and perhaps unique combination consisting of Caxton's printer's device overlying Tate's watermark was found by W.A. Churchill.[16] This would suggest a date prior to 1491 but Wynken de Worde continued to use Caxton's marks so this evidence is not conclusive.

Tate's watermark has been found in a reissue during 1494 of the Latin text of the Papal Bull of Pope Innocent VIII, to which Pope Alexander VI also agreed, expressing pleasure in the marriage of Henry Tudor and Elizabeth of York (though cousins of some degree) and in recognition of Henry as the rightful occupant of the English throne. Of the six copies extant, it is the one at Lambeth Palace which has been claimed to be the oldest piece of English paper for it shows John Tate's watermark. The other copies at St. John's College, Cambridge, Ripon Cathedral and Eton College also show Tate's watermark. This Bull was printed again as a Supplementary Proclamation in 1498.

It is reasonable to assume that paper bearing his watermark was actually made at Tate's mill near Hertford and that therefore this mill was in operation by at least 1494. While it seems that Caxton himself actually learnt how to print, John Tate would never have been able to acquire the skills of the papermaker at the vat without a long apprenticeship. He would have needed to employ skilled men, first to design and build his mill and then to run it. Such people could have been found only abroad. He needed a site where he could drive the papermaking machinery by waterpower and where there would be clean water for the actual manufacture of the paper. He would have had to arrange to lease that. After equipping the mill, he had to organise a supply of rags which may have been difficult because competitors would have been purchasing them to send abroad and people in England may not have been accustomed to save them. Then he had to hire people with the knowledge of how to make the pulp as well as the actual paper. Finally he must have had somebody who was capable of making and keeping in repair the moulds on which the paper was made. All these people must have been foreigners. Around 1800, it was estimated that at least three months would elapse between arrival of the rags at the mill and their despatch as finished paper. Therefore John Tate must have been working for quite a long time, possibly years, in order to plan, build and equip his mill before production actually started. The date of the start of Tate's schemes must have been considerably earlier than 1494, the date of the first paper from his mill of which we have any knowledge. Once the mill was running, other people had to be trained in the skills of papermaking if it were to function over a long period because this was a trade which depended upon experience to produce a good quality product.

Whatever the actual date of the foundation of Tate's mill, it must have been well established by 1495 when Tate must have been thrilled to receive the order for paper for a book, the *De Proprietatibus Rerum* by Bartholemaeus Anglicus, which had been translated by Caxton and was printed by Wynken de Worde in 1495–96. It is through a verse printed in the colophon of this book that we are able to identify Tate's paper, for it is stated:–

> 'And also of your charyte call to remembraunce
> The soule of William Caxton, first prynter of this boke
> In laten tonge at Coleyn hymself to auaunce
> That every well disposyd man may theron loke
> And John Tate the yonger, Joye mote he broke
> Whiche late hathe in England doo make this paper thynne
> That now in our Englysth this boke is prynted inne'.[17]

It was a thick volume of 478 leaves. Each leaf was folded in half so one book consumed nearly half a ream of 480 sheets of paper. It has been suggested that possibly 500 copies were printed. This paper is quite reasonable in quality, except that not all of it is equally thin, and would stand comparison with many other papers of its date. However, there are some bits of grit within the pulp, possibly coming from the water used in the vat or sand blowing into the stuff while the vatman was actually plunging his moulds into it.[18] The paper is also a trifle rusty and the sizing is a little soft.

It is in this paper that we can identify the first of Tate's watermarks which has been variously described as an eight-pointed star or spokes of a wheel within a double circle. The outer diameter of the outside circle is 1⅝₁₆ in. (33 mm.) and the inner diameter of the inner is 1⅛ in. (28 mm.). The distance across the points of the star is only ¹³⁄₁₆ in. (21 mm.) so they do not touch the surrounding circle on these early watermarks.[19] It is not known why Tate chose this watermark, but there is no reason to suppose that the design was personal to him. A motif of eight points or petals within a double circle, bearing a considerable resemblance to this watermark, was sometimes carved in Tudor wooden panelling; an example can be seen in the Tudor guest house at Topsham, Devon, which dates from the fifteenth century. The design may have been quite common at that time.[20]

The sheets of paper were formed by the vatman dipping his mould into a vat full of water with the paper fibres suspended in it. The mould was a sort of sieve, with a cover supported by a wooden framework. The vatman formed his sheets by first putting on top of the mould a deckle, a removable frame round the edge of the mould, which acted like the sides of a tray to keep the pulp on the mould. These were dipped into the vat to scoop up the pulp. The water drained through, leaving the sheet of paper on the surface. The paper in the Bartholemaeus was made on moulds constructed in two different ways. In both cases, the watermark was sewn only onto the middle of one half of the mould. In one type of construction, all the wooden ribs supporting the surface of cover were equally spaced at roughly 1⅜ in. (35 mm.) intervals apart except for the pair either side of the watermark which were at a distance of about 2 in. (50 mm.). The ribs formed shadow zones or darker areas in the paper which show up either side of the chain lines, those wires binding together all the laid wires which made up the cover itself. The laid lines were spaced at 24 per inch or 10 per 10 mm. In the first type of mould, the watermark was held in place by another chain line which was not quite central. There is no shadow zone here, so there can be no rib underneath it.

In the second type of mould, the watermark is again placed between two ribs which are further apart than most of the others at 1⅞ in. (48 mm.). In this case, the chain line supporting the watermark runs down the middle of this space. Spacing the ribs further apart at the watermark is a feature of early Italian moulds and has led to the suggestion that Tate's moulds were made perhaps by a mouldmaker from Genoa.[21] There is an additional feature in this second type of mould for the spacing of the ribs at one side of the watermark is not 1⅜ in. (35 mm.) as in the rest of the mould but only 1⁄₁₆ in. (27 mm.). So this mould must have been constructed differently from the other one and probably made a slightly smaller sheet of paper. We know the size of this mould because an example has been found in a draft court roll of Watton Woodhall Manor, written in 1500. Woodhall Manor was about four miles up-stream from Sele Mill near Hertford and both that and Sele Manor were held by the Boteler Family from the fourteenth to the eighteenth centuries. The size of this sheet is 17¾ in. (453 mm.) by 12¾ in. (325 mm.).[22] A comparison of both sets of these watermarks show sufficient differences to indicate that the paper was made at the vat with pairs of moulds, and not single moulds, which were dipped in the manner to be described later.

In 1498, Tate received a further order for paper from Wynken de Worde on which he printed three more books with the star and circle motif as the watermark and the same features in the moulds. These were *The Golden Legend* by Jacobus de Voragine with 449 leaves, *The Canterbury Tales* by Geoffrey Chaucer with 157 leaves and a much shorter work, Lydgate's *Assembly of the Gods* with 16 leaves. It is thought that only about 250 copies of these may have been printed. The character of the paper used in the Chaucer is well sized but possibly of slightly inferior quality to that produced earlier for occasionally there are lumps or small knots in the pulp and the watermark is less distinct. It has been estimated that, even allowing for trial prints and wastage, these four books would have consumed around 800 reams of paper. A paper mill, if it were to stay in business, would have to sell more than that so that Tate, after a promising start would have needed to broaden his markets.

1498 was the year when Henry VII visited the mill. At that time, Hertford Castle was a royal residence. Henry VII arrived there on 23 May and we find in the Privy Purse expenses two days later an entry of 16 s. 8 d. 'for a rewarde yven at the paper Mylne'.[23] In the following year, there was a somewhat similar entry, 'Geven a rewarde to Tate of the Mylne, 6 s. 8 d.'[24] That the paper mill was in or near Hertford is confirmed by a poem of William Vallans, printed in 1590 and called *A Tale of two Swannes*, which describes the River Lee and its tributaries. In the notes appended to the poem, there is the following,

> In the times of Henry VIII [correctly VII] there was a paper mill at Hertford
> and belonged to John Tate, whose father was Mayor of London.[25]

An ingenious reason for the second payment by Henry VII has been put forward. Wynken de Worde continued to use Tate's paper after 1500. Tate's star watermark has been traced in two more books, *Thordynary of crysten men*, published in 1506, in which a number of sheets have been used and on a single sheet in *The Justyces of peas*, published in 1510. Other watermarks appear in these books, some of which are found on the Continent, but there is one not used there, a Tudor Rose. From the way the sheets of paper are intermingled with Tate's known star, and from the quality of the paper being similar, it is almost certain that the paper was all produced at the same mill. This assumption is further strengthened by the fact that a few sheets with the Tudor Rose have been found in copies of *The Golden Legend*. The book was published originally in January 1498, and these odd sheets may represent reprinted pages to cancel an error. So it has been suggested that, after the first visit of the King in 1498, Tate's Italian workman fashioned new wire profiles emblazoned with the royal Tudor Rose and that some of the paper made with this watermark was presented later to the King himself.[26]

The moulds on which this other paper was made were different from the earlier ones, with the ribs at the watermark only 1⅝ apart (42 mm.). Once again, the locating chain line for the watermark was not central. There may have been two different pairs of moulds as the measurements vary sufficiently for the differences not to be caused by the coucher moving the mould when laying off the sheet or the unequal drying of the paper. The laid lines too on one mould seem closer together at 27 per inch (12 per 10 mm.) instead of 24. A third set of moulds may have been used with another simpler smaller variation of the Tudor Rose with an even finer cover. On this paper, while the watermark is distinct, the chain and laid lines are less so than in the other papers. The rib spacing is 1⅜ in. (35 mm.) apart but at the watermark only 1⁹⁄₁₆ in. (40 mm.). This time the watermark is in the middle between the ribs but there seems to be no supporting chain line. It has been suggested that, in *Thordynary of crysten men*, we can find a further example of Tate's watermarks, either a hand or a hand with a star above it. If so, the moulds on which this paper was made were different again.

The watermarks with Tate's known star and circle in the *Thordynary of crysten men* appear to have been made on moulds similar to the second type in the earlier books with the unequally spaced rib. However, the single sheet of Tate's paper in *The Justyces of peas* is quite different. The circle round the star is smaller at about 1³⁄₁₆ in. (30 mm.) outside diameter and the points of the star touch the inner circle. The laid lines are finer at about 35 per inch (15 to 10 mm.) but the chain lines remain thick and show clearly. The paper has quite a different feel about it. Unfortunately the evidence for yet another variation of Tate's star watermark was probably destroyed when the archives of Canterbury Cathedral were blitzed during the Second World War. In 1897, Michael Beazeley traced a star inside a circle with outside diameter 40 mm. vertically and 35 mm. across, or larger than any of the others, with the chain lines spaced differently. Beazeley gave the sheet size as 13¾ in. by 19 in. (350 x 482 mm.) and the date 1512.[27] This and the sheet in the Woodhall Manor draft are two examples of Tate's paper used for writing which points to a more widespread employment than has been considered hitherto. The variety of watermarks and moulds suggest that Tate continued to make paper for some years but for how long cannot be ascertained.

The reason for Tate's choice of Hertford probably lay in the River Lee (or Lea), because he needed transport of the raw materials, the rags, and the finished article, the paper, as well as power to work the papermaking machinery. The River Lee and its tributaries supplied both. In 1424, an Act had been passed empowering the Chancellor to issue commissions for the purpose of 'cleaning, scouring and amending the River Ley' and another Act the following year appointed commissioners.[28] There was a further Act in 1430 'To scour and amend the River Lea'. Through the work carried out under these Acts, the navigation commenced just below the Town Mills in the centre of Hertford.[29] Therefore

Tate had, for that time, good transport within easy reach of an important town and harbour, the City of London.

The River Lee itself and its tributaries were sites for many watermills. One tributary, the Beane, passes round the north of Hertford before joining the Lee just below the town. The road out of Hertford to the north follows the Beane through the manor of Ceal or Sele in which the Domesday survey records a watermill.[30] This site is within half a mile of the head of the navigation. This mill was noted in 1700 as the first in which the fine flour called Hertfordshire White was ground, and, today, the Sele Roller Flour Mills still are producing flour. Waterpower probably ceased to be used in about 1891 when the mill was rebuilt after a fire.

The deeds of 1867 conveying the Sele Mill to the forebears of the present owners, G. Garratt & Son, mention:

> Also so much of a parcel of meadowland formerly containing by admeas-
> urement one acre two roods and twenty three perches near or adjoining to
> the said last described hereditaments and being part of a certain mead lately
> occupied with the said mill formerly called paper mill mead but now or late
> called Flowers Mead bounded on one side thereof by the stream or River
> there running from the said Mill Tail and on the other side thereof by a
> stream or ditch carrying off the waste water from the said mill.[31]

A beautifully coloured plan of the mill included with this Conveyance unfortunately does not cover this meadow but it is shown as Plot 102, Flowers Mead, in the Tythe Map of 1838, drawn by W. Wilds.[32] It is the principal water meadow between the Sele Mill and the town of Hertford itself. One end forms the boundary of the mill buildings while the tailrace from the mill waterwheel runs along one edge. On the upstream side of the mill, a weir has been constructed to pen up the flow of water and create a fall to operate the waterwheel. In the 1838 Tythe Map, all this land on either side of the river is shown to belong to the 'Trustees of the late Miss Dimsdale'.[33] The riparian owners had power to control the flow of water in the river or stream adjoining their banks and the block of land owned by Miss Dimsdale's Trustees gave them the water rights to Sele Mill, with Paper Mill Mead forming a vital part of those rights. It is interesting to note that, in the fifteenth century, Sele Mill would have been far enough away from Hertford for the noise of the paper pulping stampers or hammers not to have caused a nuisance in the town.

The tradition of this meadow being called Paper Mill Mead and therefore to this being the site of Tate's mill can be traced back to two Stuart documents. The earliest surviving delimitation of the boundary of Hertford belongs to 1621 when it was set out as follows:

> From a post at the west end of the town in the road to Hertingfordbury at
> the end of Castlemead to the corner of Sealeford; then it meets the highway
> from Hertford to Watton, thence to a post near Papermill gate, and to the

northside of the river to Cowbridge, along the northside to the Lea at the east end of Hartham.[34]

Then Charles I, in the third year of his reign (1627), granted to the mayor, burgesses and commonalty of Hertford, and their successors, the soil and fishing in the waters called Benwick or Bengrade, which extended from the east end of Paper Mill Mead to a place called Goods Pool.[35] Therefore this evidence places the mill firmly to the north-west of Hertford, and not at the site of the old waterworks where L. Turnor in his map of 1830 shows 'Paper Mill Ditch'.[36]

The layout of the watercourses shown on the nineteenth century maps probably evolved after Tate's occupancy to derive the maximum power from the site. However, a visit there confirmed that the River Beane still has a reasonable flow of water. The fall at this point is insufficient for installing an overshot wheel so an undershot one must have been used. There would have been more than adequate water resources to power the mill if the type of machinery that survives today in old mills in Spain, Italy and France had been installed. The water in the river is normally clear and, coming off the chalk hills to the north, would have been suitable for papermaking.[37] Sele mill was situated quite a long way down the stream and, in times of flood, the water may have become polluted. So the site chosen by Tate would seem to have had suitable water for papermaking, adequate provision for power, good transport facilities near by, and been close to London from where its raw material, rags, would have come and where the paper would have been sold.

When production ceased at the mill is uncertain. Wynken de Worde's publication of a book with Tate's paper as late as 1510 suggests that some paper was made in the early years of that century. On the other hand, Wynken de Worde may have been finishing stocks that he had purchased earlier or even residual stocks from the mill sold on Tate's death. John Tate died in 1507 and was buried at St. Dunstan's-in-the-East, London.[38] In his will, he left to one Thomas Bolls 'as moche whit paper or other paper as shal extende to the somme of xxvi s. viii d.... oute of my paper myll at Hartford'.[39] It is interesting to note that Tate was involved with paper other than white. Either he had been importing some cheaper qualities and was using his mill as a warehouse, or he may have been manufacturing other types, presumably to keep up production because he could not sell enough white paper.

In his will, Tate directed his executors to dispose of the paper mill 'with all the goods, woods, pastures, medes, with all the commodities concerning said myll to the most advantage'. Finally, in leaving to his eldest son, Robert, all the lands in Hertfordshire and Essex, the paper mill was excluded, 'My paper myll with the appurtenaunces to be always excepted and to be sold'.[40] It is evident from this will that Tate died a wealthy man for he left money to various charities as well as his estates to his family. Why Tate failed, we shall probably never know. It may have been foreign competition, for no doubt Dutch suppliers soon found that they could reduce their prices as happens today.

Once a Bookseller made mee when I asked him why we had not white and browne paper made within the Realm as well as they had made beyonde the Sea; then he aunswered mee that there was paper made a while within the realm: at the last the man perceived that made it that he could not aforde his paper as good cheape as it came from beyond the sea, and so, he was forced to lay down the making of paper; and no blame in the man, for men will geve never the more for his paper because it was made here.[41]

It may have been difficulty in obtaining rags as was suggested soon afterwards, 'Foreigners bought up our broken linnen cloth and ragges' and sold them to us in the form of paper 'both whit and browne'.[42] Possibly demand at that time was insufficient to sustain the mill. Whatever the reason, no one else tried to start a paper mill at Hertford again.

Tate's early watermark on the mould with ribs spaced equally

Stevenson's reconstruction of the Tudor roses

1. Reconstructions of John Tate's watermarks.

12

II

PAPER OF HIGH GRADE

The Paper Maker
White linen that is torn to rags
Bought up and carried here in bags
Is washed and then to pulp is made
Then changed to paper of high grade.
JOHN LUYKEN, Haarlem 1695.

What might the mill of John Tate have been like and how would paper have been made in it? Now will follow an account of the manufacture of paper up to the fifteenth century so that the changes in the subsequent five hundred years can be understood.

Raw Materials and Their Preparation

Paper consists of cellulose fibres found almost entirely in plants. While the great majority are capable of yielding a usable fibre, the number of species which give an economically viable source are limited. In a few cases, the hairy covering around the seed provides almost pure cellulose and the cotton plant is the only important example of this type. Sometimes the fibres are grouped in a closed ring situated just below the surface of the stem and are known as bast fibres. Flax, hemp, jute and ramie and the paper mulberry are examples of these. Then there are fibres from leaves in which may be grouped esparto, manila and sisal or grasses including straw, bamboo, bagasse and maize stalks. Finally there is wood, and today both soft, coniferous, or hard, broadleaved, trees may be used. Papermaking has always needed a cheap source of cellulose fibres to enable it to compete against its rivals such as originally cloth, then parchment and today plastics. In the west, papermakers traditionally have found their basic cellulose in materials which already have been processed to remove the lignin or pithy substance surrounding the fibres in the plant. This is why linen rags were chosen because the flax had already been retted and the hard outer sheathing removed to prepare the fibres for spinning and

13

weaving. For making into textiles, the fibres may have been bleached, which, up to the beginning of the nineteenth century was done only by exposure to sunlight and sprinkling with sour milk. Subsequent wearing and washing will also have helped to soften the fibres, making them more suitable for papermaking.

Let us contrast this with the way in which the finest paper in the east was made from the best fibres of the bark of the paper mulberry tree which was one of the main sources of fibres for papermaking in China, Korea, Japan and other places in that region. The trees were regularly coppiced in the winter every couple of years or so to obtain the best fibres. The saplings were trimmed to short lengths so they could be boiled which allowed the bark to be peeled off easily. The bark was dried, really so it could be stored without deterioration and used when required. To prepare the fibres, the bark was soaked to soften it and the dark outer layer stripped off, leaving behind the long inner bast fibres. This was a tedious task, usually carried out by women. At the same time, these fibres would be washed and might be left in the sun for a few days to dry them. A further boiling followed, this time in an alkaline solution, a 'ley' made from wood ash, which broke down the lignin surrounding the fibres. Further washing followed to remove the impurities and remains of chemical treatment.[1] These stages can be found today even in the most modern methods of chemical pulp preparation.

When papermaking spread to Europe, there were no trees that yielded bast fibres comparable to the mulberry. In one of the earliest accounts of papermaking in the west, Emir Mu'izz Badis (1007–1061) gave the following description of the preparations to make white paper from hemp rope.

> Take hemp rope . . . Undo its braids and comb them with a comb until
> they are very fine and soft . . . Then prepare lime pulp, made with the best
> and whitest of quick lime, and let the hemp fibres become macerated there
> overnight until morning. Then knead them with your two hands and hang
> them out a whole day in the sun until dry. After that, return the mass to lime
> pulp, but not the same one as before, which has already served its purpose, but
> rather to a new one which you will have prepared, and leave it there overnight
> until morning. Knead it again with your hands as you did the first time and put
> it out in the sun to dry, and repeat this for three or more days. If you bathe it
> every night with a new lime pulp, the mass will come out much better.
>
> When the bleaching is finished, cut the fibres with scissors and separate
> them in pure water, which you will change daily for seven days. Then, when
> every particle of lime has disappeared, beat the fibres in a stone mortar,
> making sure that they remain fresh and moist. When they are finally clean
> and fine, when there are no dry particles or lumps, you take more water,
> making sure that the jar is totally clean, and you unravel the fibres in it
> until they look like silk.[2]

Preparation of the pulp in this way must have taken a very long time so that we can understand the advantages which the medieval papermaker had in Europe when he used rags. The importance of a supply of clean water must be noticed.

The rags had to be sorted for colour and then cut up into small pieces. This could be a dirty job, and a dangerous one, for the rags might carry plague and pestilence. Sometimes the rags might be damped and left in piles to rot by a sort of fermentation to help soften them. Lime might be added and, because heat was generated and there might be a danger of fire, the process was often carried out in stone cellars. The pleasant creamy colour of some early paper is attributed to this retting. Otherwise there was no further treatment such as boiling, for the papermakers relied upon the correct sorting and blending of the rags to give the desired colour and texture in the paper. So, in John Tate's mill, the rags would have been sorted, cut up, possibly retted and then they were ready for the next stage, that of beating.

Beating

By whatever method the pulp was prepared, there were certain changes which the fibres had to undergo before they were suitable for papermaking. These were:

1. to reduce the raw material, of whatever type and especially rags, to the individual fibres and to clean them from dirt or chemicals. This was called 'half-stuff'.
2. to shorten these fibres and fibrillate them by kneading them in water so that their outer layer turned into hydro-cellulose to produce the 'whole-stuff'.
3. to mix different kinds or batches of fibre and disperse them in water to a consistency of around 1-2% of fibres from which the paper itself could be made.

The earliest method for preparing pulp introduced into Europe by the Arabs was to crush the rags in a 'kollergang'. This machine consisted of a circular trough in which ran one, or pair of stone discs, edgerunners, set on their rims. In the simplest form, a pole through the middle of the edgerunner was fixed in a pivot in the centre of the trough and pushed round and round so that the weight of the edgerunner crushed the rags. The steep sides of the trough caused the pulp to fall back again under the edgerunner which also stirred the pulp as it went round. The width of the stone gave the fibres a sort of twisting, shearing movement which helped to fibrillate them. It is thought that the edgerunner was the first type of machine for making pulp in the early Spanish mills where it was associated with oil mills to crush the olives. In England, apples for making cider were crushed in similar troughs and examples may be seen scattered across the southern half of the country. Crushing lead ore was another application. Edgerunners, sometimes with one but more often with two stones, have been a feature of some paper mills right up to the present day. At Amalfi in Italy, a few small papermills still make their pulp with edgerunners which may reflect a continuous tradition. In England, this machine does not seem to have been used until it was introduced to the Nash Mill of John Dickinson & Co. in 1880. Edgerunners became popular for pulping waste paper but are now outmoded and scrapped.[3]

The more usual method of preparing pulp in medieval times was with the stamper or beater. Beating originated with the pre-paper substances of papyrus and 'Tapa' or bark-cloth. To make papyrus, the reed, about eight to ten feet high, was harvested from the Nile and the ends trimmed off, leaving the central section. The outer sheathing was peeled off and the inner pith cut into thin strips. The strips were soaked in water and today are rolled to compact them. They were placed side by side on a board in two layers, one at right angles to the other, to form the sheet. These layers were hammered and pressed together before being dried as a sheet of papyrus.

In a similar way, to make bark cloth, the mulberry fibres, which had been prepared as described previously, might be laid out side by side and across each other and then made to cohere by hammering with hand-held wooden mallets. A little water had to be added to keep the sheet plastic. The production rate might be one or two sheets a day. Another way was to take a mass of fibres and beat them out into a mat, rather like rolling out pastry, with water being sprinkled on them to keep them moist. Mulberry fibres for making true paper were beaten in the same way. The lump of fibres was beaten on a broad board or flat stone with oak sticks or mallets. The faces of the mallets might be carved into sharper points to give a cutting action or blunter ones to fibrillate the fibres. A simpler method of beating would be hard to devise and in places in Japan, Korea and Nepal, it has survived up to the present day.

In China, pulp made from shorter fibre stock, such as bamboo or rags, almost certainly was beaten in hand- and later feet-operated pestles and mortars. It is known that the pestle and mortar have been used to prepare cereals for human consumption, such as 'hulling' rice or crushing oats, and the origins are lost in the mists of antiquity.[4] Usually, only a single person with one pestle was involved but sometimes there were two with two pestles. Some pictures purporting to illustrate early papermaking in Europe show the same method. Some heavy pestles for crushing corn were suspended above their mortars from spring poles or branches of trees which helped to relieve their weight if not the monotony of this tedious work. Mortars with this addition were used in North America and have been seen in Yugoslavia after the Second World War.[5] However, it is unlikely that Tate would have made his pulp in such a manner because he built a waterpowered mill which suggests a battery of stampers.

Another way of hulling rice, which again has been used since time immemorial in many parts of Asia including China, was with the treadle or 'tail' stamper. It was adapted to prepare pulp for papermaking in these regions too, a process which is being carried on up to the present day in parts of Thailand, India and Nepal. These stampers have remained unaltered for centuries. For papermaking, one, but more often two people, pressed down with one foot each on the end of a pivoted shaft so that they raised the other end which was fitted with a hammer head. Lt. Col. W. Ironside saw such primitive 'tail hammers' being worked to pulp rags in the Hindustan, India, before 1774 and gave a description.

a) A Stamping lever, ten feet long, and seven inches squared timber.
b) Two pieces of wood, fixed in the floor, to support the axis of the lever.
c) This end of the lever is pressed down by the feet of two men.
d) is a stick, suspended from the roof of the house, to which are fastened four ropes which support the arms of the workmen.
e) The head of the stamper four feet long, and four inches square timber bound and shod with iron.
f) a perpendicular section of terrassed cistern, dug in the ground floor about 4 or 5 feet square.
g) A square stone, on the bottom of the cistern, excavated in the middle, to receive the head of the stamper, by which the rags are beat to pieces. A person is stationed in the cistern, to supply the stamper with rags.[6]

Ironside said that, for the best quality white paper, the rags or pulp were alternately beaten in the stamper and then left to bleach in the sun for a few days. This could be repeated seven or eight times. It should be noted that the rags were beaten in a moist state, like the mulberry fibres, but were not floating in water. This was why it was necessary to have a person squatting at the hammer head to push the rags or pulp into a hole below the beating face of the hammer.

When J. Trier visited Nepal in the 1960s, he found one place where this style of stamper was still operated by foot power. Two men trod alternately to give a rate of about 25 beats per minute. A woman at the other end of the beam pushed the escaping paper slurry back into the hole with a wooden stick. The work was both laborious and time-consuming, for it took three people 6 to 8 hours to prepare 10 lbs. (5 kilos) of pulp.[7] In Kashmir, the driving system was improved over the foot-operated stamper because cams on the axleshaft of a primitive waterwheel pushed down the tail of the stamper. This dispensed with the two people who provided the foot-power while leaving plenty of room round the hammer head for a person to feed the rags under the hammer head.

The pulp in the early Moorish style papers, which was the first type of paper made in both Spain and Italy, was poorly beaten and lumpy. Short lengths of twisted yarns and curls of partly beaten groups of fibres can be detected, while the paper is thick and difficult to see through. Where it become worn or rubbed at the edges, it looks rather like cotton wool as it has such long fibres. Very soon after papermaking was introduced to Spain and Italy, it became associated with wool merchants. Such people had vital international trading links and well developed manufacturing techniques, particularly in the Tuscany region of Italy. At this period the wool trade was flourishing and could have provided the capital to invest in a new industry. Then there was a transfer of technology and techniques from the one industry to the other. Presumably the papermakers needed 'felts' which were made from woven woollen cloth to separate the wet sheets of newly formed paper. The rags for papermaking could have been cut up with scissors either like those for shearing sheep or possibly like the large shears for cutting the nap on woollen cloth. Vegetable starch was replaced by gelatine from animals to size the paper. Another link was the fulling mill, originally used for scouring and felting the woollen cloth but soon adapted for pulping rags for paper. It was probably on the fulling mill that the earliest papermaking stampers were based in Europe. Water-powered stampers are thought to have been operating at Valencia in 1151 A.D. and so it is likely that Italy, following the Spanish lead, had early waterpowered mills too.

During the fourteenth century, a new type of paper evolved in Italy. It is associated with the art of watermarking for watermarks would not show clearly in the Moorish style of paper. 1282 is the traditional date for the introduction of watermarking which first appeared at Fabriano in the centre of Italy. The first watermarks are difficult to detect clearly in the very early Italian papers which also are poorly beaten and lumpy. Yet in the next hundred years there must have been a change in beating techniques for the quality of the paper steadily improved and, by 1380, the Italians were making paper that stands comparison with modern production, well beaten, even consistency and clear look-through, so that the watermarks and mould construction show up clearly. This paper was sized with gelatine and was much harder. This new Italian paper had reached a sufficient quality for it to be exported by the middle of the fourteenth century and the Spanish manufacturers had to change their production methods to compete.[8] Obviously there had been a technical revolution, but just what had happened we can surmise only by looking at much later apparatus. However, these Italian techniques were the ones which spread across the rest of Europe and formed the basis from which our present industry grew.

The paper industry arrived in Italy when fulling stocks for washing and felting cloth were probably well developed and some features on them may have been adapted to improve the papermaking stampers. On fulling stocks, the hammers were raised at their heads and not their tails as in eastern stampers. Fulling stocks had two hammers pounding the cloth in a trough where the cloth could be soaking in the fulling liquor and where the cloth could be turned round and round by the action of the hammers themselves, features which may have been taken over into papermaking to improve the beating. Fulling stocks have been preserved which may show how they originated and how stampers evolved out of them. In Rumania, at both the Open Air Museum in Bucharest and in another one beside Count Dracula's Castle, fulling stocks have been seen consisting simply of two pairs of straight hammer heads with pegs or tappets sticking out near the middle of their backs by which they could be lifted by cams on a waterwheel axle. These heads were inclined away from the trough over the camshaft and slid up and down on wooden guides. They were almost like a couple of large pestles pounding the cloth in the

2. A set of waterpowered stampers at Amalfi, Italy. There are three troughs with three hammers per trough on this side of the partly encased waterwheel and another trough beyond.

trough below. The feet were cut back in steps to help rotate the cloth and pound it more evenly. Access to the trough was easy because there was no obstruction from pivot shafts but the heads, sliding on the guides, must have caused friction even if they were lubricated with splashes of water. It might be noted that in some Dutch windmills there were fulling stocks with vertical hammers lifted by camshafts placed at the tops of the heads of the hammers. There was a similar arrangement in Dutch papermaking windmills, such as the Schoolmaster, where vertical hammers with knives on their lower ends cut up the rags.[9]

Another type of fulling stock seen in Rumania could have been derived from the previous ones and the hand pestles with spring beams above them. These stocks had pivot shafts fixed in the top of the heads of the hammers from where the shafts sloped upwards to pivots fixed high in a frame suspended from the roof of the shed. The waterwheel axle with the cams on it was placed near the floor on the far side of the hammers and troughs. This meant that the lifting tappets had to be made from extra pieces of wood stuck in the middle of the hammers resembling the method of the earlier stocks described above. Access to the troughs was easy under the pivot shafts but, structurally, the layout seemed weak and occupied a great deal of space.

In a third type of Rumanian fulling stock, the pivot shaft was set much lower and ran, still sloping down, straight through the head to form the lifting tappet with the camshaft at the far side. The pivots were supported by a frame built on the floor, so much less space was required but access to the troughs was more difficult. The hammers were still guided by framing at the sides of each pair of heads, which was the only structural difference between the fulling stocks in Rumania and paper stampers in Italy at Fabriano and Pescia and in Spain at Capellades. One technical difference between the stocks and stampers was the inclination of the hammer head on the pivot shafts. In fulling stocks, these point towards the pivot while on stampers they point outwards. Yet on the stampers at Amalfi, the heads pointed inwards and the shafts incline upwards. In spite of this, the similarity of early fulling stocks and papermaking stampers has now become apparent and suggests the origin of stampers used in the west.

This possible derivation might account for the first difference between the eastern and western stampers, where the lifting cams were placed at opposite ends of the pivot shafts. A second difference was that, in the west, the surviving remains and illustrations show that the rags were beaten with more than one hammer (usually three in southern Europe) in each beating trough. The origin of this feature has been ascribed to thirteenth century papermaking at Fabriano[10] and once again suggests links with fulling stocks. A third difference was that, in the west, the rags were beaten in troughs filled with water which effected a radical change in pulp preparation. The difficulty of feeding rags into three hammer heads pounding consecutively, or the more difficult access caused by the layout of stampers based on the last design of fulling stocks, may have been reasons for substituting a trough for the earlier board or stone. It meant that the beating process no longer needed to have the constant attention of a person sitting at the head of the hammer to feed in the rags, as was necessary in the east. So the simple addition of a water-filled trough must have helped to cut production costs by dispensing with a tedious labouring job as well as having the additional advantage of preparing a more even pulp.

There was almost certainly a fourth difference between the eastern and western stampers, for, instead of the one hammer in a single trough to which the rags were returned time and time again, the surviving sets of stampers in Spain and France show that the pulp was passed from one trough to another in which it was beaten by hammers faced with different types of feet. Once again, this change is ascribed to Italy,[11] for it is thought that here the stamper feet were first fitted with spikes to shred and pound the rags. Unfortunately the surviving Italian sets have been modified to work in conjunction with the later Hollander roll beaters and now all have the same type of nails.

In the old Spanish paper mills at Capellades and in the valley above La Riba, the stamper troughs were set in groups of three. They have been carved out of the local travertine stone in such a way that the inner walls were curved to deflect splashes of pulp back into the troughs. In the bottom, iron or bronze bedplates were set in wood. At other papermaking sites, suitable materials available locally would be used, which in northern Europe often meant wood. In Spain, the rags were placed in the first trough where the hammer heads were nailed with spikes that tapered into a cutting edge at the bottom. There were five rows with five spikes in each. Here the rags were chopped up against the iron bedplate, a process which continued from 12 to 18 hours to turn the rags into 'half-stuff'. In the next trough, the feet of the stampers had nails with triple-ribbed heads, again set five by five. Here the half-stuff was beaten rather than cut for about twelve hours until the pulp became 'whole-stuff'. In the third trough, the feet of the hammers were left plain wood so in this trough occurred a final beating or blending of different pulps and it would have been used for a much shorter time. At Capellades, a furnish for the first trough was 12–17 kilos of rags.

Another important feature of these stampers was that clean water could be run into them and let out through a hair screen or filter so that the rags and fibres were washed as they were beaten. This was unnecessary in the final trough. The water was poured in at one corner and let out at the opposite side to assist in the circulation of the pulp. Circulation was also assisted by the angle of the hammer heads which were not set vertically and by the angle of the pivot shaft which caused the heads to have a sideways as well as a vertical movement. This could dramatically affect the volume swept by the heads which might range from 8 per cent at Arnhem to 39 per cent at Pescia. Then the hammers were raised in the sequence, 1, 2, 3, away from the point of entry of the water in order to beat the pulp evenly. The rotation of the camshaft ranged from 20 to 30 r.p.m. so with two trips per rotation, the number of the blows was double this.[12]

This would have been essentially the beating system used by John Tate. Minor differences appeared as papermaking spread northwards. The number of troughs might vary depending upon the water power available. In France, we find sets of six troughs and in Holland the usual number was four. These would be divided between the different stages. Another small difference between most of the stampers south of the Alps and Pyreneese and those to the north is that the angle of the pivot shafts in the north slope up from the back pivot through the hammer heads to the camshaft. The method of nailing might vary from region to region with much larger ones being used in Holland. In Germany and Holland, there were often four, usually five and sometimes six hammers to each trough.[13] Huxham Mill in Devon had five hammers per trough. The heads of these hammers were made from a smaller section of wood and so presumably were lighter than their southern counterparts. Unfortunately no stampers have survived in Britain and there are no drawings either, so we remain ignorant of their layout and design in British mills. However, the fact that stampers were an essential part of the paper manufacturing process meant that early mills had to be located on streams where there was an adequate supply of water to provide power. This was one factor determining the localization of the paper industry as it expanded in Britain.

Forming the Sheet

The pulp, or 'stuff', to give it its correct name, was now ready for actually making paper and might be kept in a storage chest ready for tipping into the vat from which the sheets of paper were formed. In the most ancient Chinese method of papermaking, the sheets were formed in a floating mould. In any convenient still pond or tub of water was floated a wooden frame with a loosely woven material stretched across the bottom. The water filled the bottom of the mould. A measured amount of pulp, sufficient to give the correct thickness or weight of the final sheet, was poured into the mould and dispersed evenly either by the fingers or by a stick. As the mould was gently raised horizontally, the water drained out through the bottom cloth, leaving behind the fibres across it. The whole mould was stood in the sun or around a fire and the sheet of paper pulled off when dry. This necessitated a large number of moulds and production was limited to 50 to 100 sheets a day.

In the east, there was another method of forming the sheet. A large tub or vat was filled with the stuff which was diluted to a working consistency of 1–2% fibre content. While more pulp was needed, techniques were developed which considerably increased production. The mould was dipped into the vat and lifted out so the sheet of paper was formed on top of it and the water could drain away through the cover. The mould consisted of a frame with a series of parallel ribs, pointed at their top edges, stretching across between two opposite sides. The tops of the ribs and the tops of the side framing were all at the same level so they could all support the cover which was placed on top. In eastern papermaking, this cover was made in a special way from thin strips of bamboo or grass placed parallel to each other from left to right. They were sewn up into a sort of mat by fine thread or horse hair from top to bottom (or back to front). At the top and bottom of the cover, the binding threads were tied round larger pieces of bamboo or stick which strengthened the cover and also formed two of the edges of the mould. In the finest moulds, the bamboo strips might be less than 1 mm. thick and were spaced at the same distance apart. The water could run through the spaces leaving the fibres behind. More fibres tended to accumulate in the spaces so the paper became thicker there while on top of the strips, the paper would be thinner. This shows up as a pattern of darker and lighter lines across the paper which are called 'laid' lines. The binding threads may also appear as thin light lines running at right angles to the laid lines and they are called 'chain'

22

lines. This type of cover has good draining characteristics and has continued in use right up to the present day.

In the east, the removable cover was placed on the framing of the mould and held there by a pair of 'deckle' sticks at either side which completed the edging around the top with the two sticks sewn into the ends of the cover. They made a sort of tray which could be filled with pulp in different ways. The mould could be used in the same way as the pouring moulds. But in Kashmir, the pulp was prepared in a vat, stirred and the larger lumps allowed to settle. The cover was placed on the mould and both carefully dipped and kept under the surface until covered with fibres as they settled. Both were lifted out and stood up to drain. This dipping might be repeated two or three times, depending upon the thickness that was wanted in the sheet. As can be imagined, this was a very slow process for it took about four minutes to form and couch a sheet. About 250 sheets were produced in a day.

Yet another way was to put the cover on the framework of the mould and, holding it there with the deckle sticks, the vatman could scoop up some stuff on the surface, swish it around until it was even across the whole of the mould and then let the remainder of the water drain out. This was of course much quicker than the previous method and is still employed today by many eastern papermakers. Production by this method has reached 1,500 to 2,000 sheets a day.

In both cases where the mould with the flexible cover was used, the cover with the wet sheet on it was lifted off the framing, turned over and the damp sheet laid on top either of a board or on top of the one previously made as the pile was built up. The cover was peeled or rolled up off the sheet so it had to be flexible. In Kashmir, a pile of 140–170 sheets was built up before being pressed to squeeze out more water. In Japan, a thin thread was laid across one side of each sheet which was used to separate the sheets as they were pulled off one by one and smoothed out on a plastered wall to dry. In some types of paper, especially that made with mulberry fibres, a mucilaginous liquid, derived from the root of the Takpul plant, was added to the vat which helped in the dispersal of the fibres and also prevented the sheets of damp paper from sticking together.[14] When dry, the sheets fell off and apparently it was quite common in Kashmir to see men running down the street chasing paper which had been blown off by the wind.

It is probable that this was the type of mould on which the early Spanish and Italian papers were made. What is certain is that their early moulds were unlike those constructed later. This paper varies in quality. In some sheets, the paper is rough and poor, but in others, the raw materials must have been chosen with care to give a good white paper but all this paper is thick and heavy because the fibres are long. The paper is not watermarked and even the laid and chain lines are difficult to identify. There are probably between 12 to 15 laid lines in 20 mm. on a typical sheet and they will be rather irregular. The chain lines are spaced at one or other of two intervals, around 1⅜–1½ in. (36–8 mm.) or 2 in. (50 mm.). They are not always parallel, and in some sheets are distinctly wavy. It has not been possible to determine with any accuracy whether there are any stitching marks to show whether the chain lines and so the cover itself has been secured to the mould framing. The look of the paper suggests there are none. This lack of evidence, together with the wavy chain lines and the lack of shadow zones to indicate where the ribs may have been, point to the use of a mould with a removable cover, although the lack of shadow zones could be due to the long fibred pulp or possibly the dipping procedure.

In Europe, there was no suitable material like bamboo from which to make the laid strips for the cover and it is thought that, in Italy, copper wire soon replaced any earlier type of material. During the fourteenth century, watermarks in the paper show that the wire from which the covers were made became thinner, as if the technique of making the wire was refined. Also it seems that the covers must have been sewn onto the framing of the moulds. These were among the developments introduced by the Italians which helped to speed up production as well as improve the quality of their paper.

In the earliest sheets of Italian paper, including some as late as 1343 and 1344, the laid lines are very irregular, indicating not only that they were not straight when they were woven into the cover but that their diameter may not have been constant. One suggestion is that copper sheet was cut into strips and hammered into round wire. Wire-drawers are mentioned in Augsburg, Germany, in 1351 but the art was known earlier in Italy. Around 1100, the monk Theophilus described wire-drawing so drawn wire ought to have been available for making covers and watermarks but possibly at first not of very good quality. Mechanical wire-drawing was invented around 1400.[15] The chain lines in this Italian paper are very weak and thin so they still may be sewn with hair. Then in 1347 appears a sheet watermarked with a horse. The laid lines are strong, straight and regular, 3 mm. apart. Shadow zones have appeared, the watermark is distinct, but the chain lines are still weak. It looks as if there has been a sudden improvement in the art of wire-drawing, for the wire has become not only straight but also constant in diameter. These characteristics are maintained in later moulds with the exception of the chain lines.

It is not until 1385 that the laid lines become much closer together, at approximately 1.8 mm., suggesting further improvements in wire-drawing. Also the chain lines are heavier, possibly indicating that thin wire now bound the laid lines, and distinct marks appear in the paper indicating that the cover was sewn onto the mould. On these papers, the ribs are situated underneath the chain lines and light marks appear on the chain lines at regular intervals along them which must be stitches sewing them onto the ribs. At first, the same laid line right across the mould was sewn to the ribs but it was found that the cover sagged and stretched between the stitches so on later moulds the stitches are staggered. Certainly by the end of the fourteenth century, fine wire for binding and sewing must have become available. The better quality of the wire and improved pulp in 1397 produced a beautiful sheet of paper with a clear, distinct watermark of a dagger. The paper itself is remarkably white with even beating and clean look-through and with distinct shadow zones. The laid lines are straight and parallel, about 1.2 mm. apart, the chain lines being 35 mm. apart. Stitches can be seen on the chain lines. This sheet has every characteristic found in the best sheets of hand-made paper ever since. In fact, we can say that by this date, the traditional western system of paper manufacture must have evolved and must have been similar to the way John Tate produced his paper.

Western Techniques of Papermaking

We can say that the moulds on which John Tate made his paper had a wire covering firmly sewn onto them so we must examine how the paper was made. There has survived a mill at Ambert in France where paper is still made by hand in a way that may have altered little since the Middle Ages. First a vat of stuff was prepared by the beaterman emptying enough pulp into it to make a complete 'post' of paper. A post normally consisted of 144 sheets but it depended upon the type of paper and how many sheets would fit into the press.[16] The stuff in the vat was stirred up with a paddle to mix and disperse the fibres throughout the water. The vat might be topped up with water if necessary but nothing else added until that post was finished. This meant that, as the vatman formed each sheet and took fibres out of the vat, the stuff would have become thinner and therefore he must have compensated for this to keep the sheets the same thickness.

When making large quantities of ordinary paper, the vatman worked with a pair of moulds and one deckle. When the vatman was ready to make a post, he gave the vat a further stir with the paddle. Then he took a mould, put the deckle on top and scooped up some pulp. He immersed only about half of the mould. He lifted it out, levelling it and spreading the pulp across it. He tipped some back into the vat over the far edge and so controlled the thickness of the sheet. At this point, the far edge of the mould might touch the stuff in the vat but the mould was never totally immersed. The mould was shaken with a special papermaker's 'shake' to send a ripple of stuff across it in both directions to interlock the fibres and 'close' the sheet. The vatman rested the mould on the side of the vat to allow it to drain. The vatman paused, and then stirred the stuff in the vat with his right hand.

While one sheet was being formed, the second member of the team, the coucher, was couching the earlier sheet made on the second mould onto a damp woven woollen cloth called a felt. Unlike papermaking in the east, the newly formed sheet of paper was far too wet and soft to be placed directly on an earlier one for they would have stuck together like a lump of cardboard. These felts were another link with the wool merchants and had a raised surface like a strong blanket. When the coucher had deposited the paper on the felt, he pushed the empty mould along a plank or bridge placed across the vat to where the vatman would be able to reach it.

The vatman then passed his mould across to the coucher, lifting off the deckle with both hands in such a way that no water would fall off onto the new sheet of paper. The coucher took this mould and lent it against a small post, the asp, to drain. The vatman, meanwhile, was reaching for the second, empty mould to begin again. While the vatman was dipping, the coucher was laying another felt on top of the sheet of paper he had couched off previously. While the vatman was stirring the stuff in the vat, the coucher couched the sheet. At the bottom of the post, the vatman rocked the mould from side to side a couple of times to remove the sheet, but, as the post built up, he rocked it only once. Just when this routine with two moulds was evolved is unknown but the watermarks in Tate's paper show that it was established in his day. Sometimes today when a paper has a very complicated watermark, only one mould may be used but then the output is much slower.

Pressing, Drying and Sizing

When the correct number of sheets had been made, they were dragged under a massive wooden screw press. Such presses may have originated with wine presses but a much more likely source could have been the cloth press found in the wool industry. Blocks of wood might be placed on top of the post of felts and paper to level them so that the pressure would be applied equally when the head was screwed down. The vat team swung on the end of a long lever to tighten the screw and to squeeze out as much water as they could. Tradition has it that they then blew a horn to summon other people from the rest of the mill. A rope from the samson post, a form of winch, was tied round the end of the lever and every body swung on the arms of the samson post to give maximum pressure. A ratchet prevented the screw on the press turning backwards. The free edges of the felts were scraped to remove the last drops of water and then the press could be released and the post pulled out.

Pictures show a third member of the vat team, the layer, separating the felts and the sheets of paper. The paper was placed on a sloping stool while the felts were put ready for the coucher to use again. It was claimed in 1816 that it was a day's work for three men to manufacture four thousand small sheets of paper.[17] The pack or wad of wet sheets was carried to the top of the mill where the drying lofts were situated. The design of these lofts must have been dictated by local conditions and local building materials. In the south of Europe, they are part of the normal stone structure of the building. In the Ambert region of France, they are constructed from wood on top of the stone lower floors. In Germany, where the depth of snow in winter necessitates a steeply pitched roof, they are contained within the eyes. No pictures of John Tate's mill have survived to show us what style he may have followed. All these drying lofts had common features, for all had shutters to control the air blowing through them to dry the paper and all had ropes, like washing lines, over which the paper was hung to dry.

3. Vatman dipping a mould into the vat at Hayle Mill, Maidstone, 1969. On this occasion, a paper of double thickness was being formed by couching a second sheet on top of the first so, in the background, a third mould was being used to press out more water.

The person hanging up the paper picked up the corner of a sheet from the damp pack and placed a papermaker's cross under the middle. The cross was shaped like a letter T and the person held the lower part in his other hand so the paper rested over the crossbar. Then with a sweep of his arm, he swung the paper up to the rope and left it hanging down either side. The paper would be returned to the drying loft for a second time after it had been sized. This time, much greater care in drying was required to prevent the size spoiling and discolouring.

The early Spanish paper can have a wonderful translucency caused by its size. A vegetable size was made from either rice or wheat starch. Sizing techniques in Kashmir show that the starch paste was spread over the surface of the paper, rather like applying butter to bread. It was then smoothed and polished by

4. A small paper mill at Winchcombe, Gloucestershire, showing the drying loft on the top floor in 1978 before conversion into a private dwelling. The stream that powered it still flows at the foot of the gardens.

rubbing a pebble across the surface. It appears to have been applied more thickly than later sizes and it is this which both gives the smoothness of the surface but also has tended to obscure the laid and chain lines. The original type of Spanish paper appears to have been sized in this way until its manufacture ceased around 1380.

The later Italian paper has quite a different feel and appearance from the Spanish which was partly due to the different sizing techniques. From about 1300, Italian size was probably made from gelatine, obtained by boiling the bones, skins and hooves of animals; another link with the wool industry. In paper of about that date, the sizing appears streaky, as if, possibly, it had been brushed on and then polished. Even so, this paper has a crisper feel than Spanish sized with starch. However, by about 1370, Italian sizing becomes more uniform and this improvement is maintained in later papers. This effect might have been achieved either by pouring the size over the sheets or by dipping them into a tub of size. When dry, polishing by hand through rubbing a stone, such as an agate, across the surface would still have been necessary.

At last the sheets of paper were finished. To achieve paper of high grade, John Tate must have constructed all this equipment and established all these manufacturing stages in his Sele Mill at Hertford before he could start making paper in the first papermill in England.

III

THE ART OF
WATERMARKING

A definition of a watermark is a contrived thickening or thinning in a sheet of paper so that a predetermined design is produced at the same time as the paper is being formed and is an inherent part of the paper itself. The surface of the mould is arranged so that some parts of it may be higher and other parts lower than the ambient level. When the paper is held up to the light, the parts of the mould which were higher show through as lighter areas, while the lower parts become darker in appearance. The pulp forms around these patterns and therefore, for clear watermarks, the fibres must be short, but then the paper tends to become weak, especially at the thinner places. There is also the problem of distortion when drying a sheet of paper with different thicknesses and also when the fibres may not lie equally in all directions. The sheet may contract at a different rate in its width or length, a problem encountered particularly on paper machines, and allowance may have to be made for these problems when designing the watermark. Why these differences in thickness are called watermarks is not known, as they are not caused by water to any greater extent than on any other part of the sheet. The English term 'watermark' is confusing and the French 'filigrane' or the Dutch 'papiermerken' are more appropriate. It was to be many centuries before moulds were made with parts of their surfaces deliberately lower, but the light watermark had its origin at Fabriano in Italy in about 1282.

There is no evidence that any watermarking techniques were developed in the east until western ideas were copied about one hundred years ago with patterns of ramie grass sewn on the surface of a flexible cover.[1] In the west, the earliest watermarks have quite simple shapes, such as the outline of the head, shoulders and arms of a person, a letter like a 'P', a cross, circles, a bull's head, a dagger and so on. The assumption is that they were bent from lengths of wire and sewn onto the surface of the mould. While it is possible that grass or thin wood strips might have been used, some of the corners seem too acute and the curves too sharp to have been bent from

5. A 'Britannia' watermark sewn onto a laid cover.

any material other than wire. When couching with the removable eastern cover, the cover itself was flexed and even rolled up as the sheet was couched off it. While wire has a certain amount of flexibility, it is doubtful if it has enough to couch in the eastern style. Also it is doubtful if the stitches binding the watermark could have withstood this sort of treatment for long. Therefore the introduction of watermarked paper at Fabriano in the 1280s must mark the development of the western style mould in which the cover was sewn firmly onto the ribs. This also necessitated the introduction of the removable deckle, for otherwise the sheet could not be couched.

What was the purpose of a watermark? The mark could identify who made or commissioned the manufacture of that sheet. Each mill would have its own special marks which would show not only who had manufactured the paper but also the quality. Therefore watermarks were a guarantee of standards too. Watermarks would be varied from time to time so, theoretically, it ought to be possible to determine the origin and broad dating of a piece of paper from its watermark. Because watermarks are formed in the paper during the actual manufacture, it ought not to be possible to counterfeit them. Therefore watermarked paper can have a uniqueness which has given it a crucial importance as the material on which can be printed bank notes and other documents where security is essential. Of course unscrupulous rivals have always tried to copy the watermarks of other famous manufacturers and, to prevent this, watermarks of greater and greater complexity were evolved gradually. Here it might be recorded that, during the French Revolution, from 1793 to 1795, the English government counterfeited Assignats of various values at Haughton Mill near Hexham in Northumberland.[2] This mill is still standing, hidden away in the hills to the north-west of that town.

The position of early watermarks was nearly always in the middle of the sheet or in the middle of each half sheet. The wire profiles were sewn onto the laid lines which caused a defect because they could slide along the wires during couching. Occasionally it is possible to find a series of sheets of paper made consecutively from the same vat where this movement of the watermark can be observed as the gap between the watermark and one chain line widens at the same time as it diminishes on the other. The solution to this problem is found in later sheets where the watermark is positioned with a chain line running through the middle of it. As watermarks grew more complex and bigger, they stretched across chain lines so that the problem of their movement became less acute. However, it became traditional to arrange them so that they were centred around chain lines and the layout was balanced.

In some early papers, it can be observed that the chain lines are more widely spaced at the watermark than across the rest of the sheet. In a sheet of paper dated to 1348[3] the ordinary chain lines are spaced at 42 mm. apart except for those on either side of the watermark of a cock which are 54 mm. What appears to be a chain line through the middle of the watermark shows up thinner than the others and has no shadow zone beneath it. In many sheets, this peculiar chain line may often wander from side to side as if it is less well attached than the others. It has no marks indicating that it was stitched to the framing of the mould and so it would appear to be a sort of binding wire to strengthen the mould at this point, where probably the ribs were spaced more widely, as well as to help secure the watermark.

In nearly all laid paper made on the western type of mould before 1800, it is possible to see thicker regions of pulp running across the sheets at right angles to the laid lines. They have been caused by the ribs of the mould framing beneath the cover. The ribs support the cover to prevent it from sagging when the vatman dips the mould and forms the sheet. They also support and strengthen the cover when the coucher turns the mould upside down and couches off the sheet. Because the cover is sewn to the ribs, the cover will not fall away from the framing as the coucher turns the mould upside down. The ribs also perform another vital function in draining the water from the pulp, which may have passed unrecognised for a long time. People who have been camping soon learn not to touch the sides of their tents when it is raining, otherwise the water will penetrate at that point. In a similar way, where the mould cover touches the ribs, the water is drawn by surface tension and runs away down them. But this also draws some of the pulp to the ribs so that the paper becomes a little thicker at those places where they touch the cover, thus forming the shadow zones.

The shadow zones show that the cover was made from a single layer of wire and reveal where the ribs were situated for they will be formed only on either side of the ribs. This may also show how the cover was sewn on. Sometimes, particularly in later Spanish and Swedish papers, the shadow zones may be seen to one side of the chain lines and occasionally it seems as if there is a second chain line running down the middle of them. Existing Spanish moulds show that the proper chain lines have been arranged to one side of the ribs, presumably because it was more difficult to sew them on top of the pointed ribs. So they do not have the shadow zone beneath them. The second chain line must therefore be a false one, either caused by the rib itself protruding through the laid lines or pulp building up on the rib in a dense lump and so causing the double effect. The latter reason is probably correct because Johann Krunitz wrote in 1807,

> 'If the moulds are dirty, then they must be thoroughly cleaned; because the
> fine particles of the pulp most frequently deposit in some corners of the wires
> or watermark, so as to cause white irregular stripes'.[4]

The custom later in England was for the chain lines always to be sewn directly along the tops of the ribs so that the shadow zones form equally either side of them. In English moulds, the ribs too become equally spaced across the whole of the mould at roughly 27 mm. intervals, but this will vary depending upon the manufacturer and the type of paper.

Now the laid and chain lines, the watermarks and the shadow zones all caused the paper to vary in thickness. The finer the cover of the mould, the smoother the paper but the inequalities persisted. These variations presented great difficulties to printers who were trying to reproduce fine details in copper engravings or were using very thin type faces. In the middle of the eighteenth century, John Baskerville, a famous printer in Birmingham, designed a typeface with very narrow lines and found difficulty in printing with it on the laid paper then available. The full account of how he solved his problem with the help of James Whatman the elder will be told later. Here it will suffice to say that in 1757 Baskerville created papermaking history by publishing his quarto edition of *Publii Virgilii Maronis Bucolica, Georgica, et Aeneis*, the first book in the western world partly printed on wove paper. These wove sheets have no watermarks at all and must have been formed on a mould with a cover of wire woven like a piece of cloth, in other words, on a cover similar to that of the earliest Chinese floating moulds. However it took some while before all the problems in making this type of paper were resolved and perfect wove paper appeared. Wove covers took longer to drain than laid ones so were unpopular with the vatman but the papermakers could charge a higher price for paper produced on them.

The invention of the wove cover was crucial for the papermaking industry in many ways. First it led to a better quality product with smoother surface and more uniform thickness. This was important not only for letterpress printing in books but also for printing copper engravings and for notepaper as well as artist's watercolour paper. When writing on laid paper, a quill pen might catch in the furrows left by the laid wires, and these hollows might also trap pools of water in painting.[5] Then, because the cover was constructed from warp and weft wires which interlaced with each other in a series of holes, it was easier to sew on complex watermarks so they were immovable. This would become important for the manufacture of security papers. Indeed in the early nineteenth century, it was realised that wove wire could be moulded in ridges and hollows to give watermarks with graded or three-dimensional effects which have become vital in security papers. Although probably used on the earlier experimental papermaking machines, John Gamble and the Fourdriniers definitely mention in their patent of 1807 that their improved machine had wove wire. The Fourdrinier machine would have been almost impossible to operate successfully with laid wire.[6] So the elder James Whatman's introduction of wove wire around 1754 was to have far reaching effects.

During the eighteenth century, designs of watermarks steadily increased in complexity. These presented all sorts of problems to the mould maker as the designs had to be assembled from bent pieces of wire. Possibly he may have worked from a pencilled sketch and bent the wires with his fingers. Pliers with flat or circular noses and other implements might be adapted by the ingenuity of the craftsman to bend the necessary shapes. Later, the design might be carved into a wooden board and the wire profile formed to that shape. This gave the possibility of duplicating designs more accurately.[7] The ends of the wires were particularly vulnerable and

6. A complete watermark made with an ordinary laid cover around the edge, wavy laid in the centre and the lettering pressed out of wove wire.

might become loose, perhaps causing holes in the paper. Looking through sheets made consecutively, it is sometimes possible to see how some of the sewing wires have broken in a watermark, allowing that part to become deformed and finally to fall off. Therefore, the fewer pieces of wire with which the mould maker could fashion his profiles, the stronger would be the wire profile.

If one wire crossed another in a wire profile, the height of the upper wire would be twice that of the rest of the design and would show as a lighter spot which also might form a hole. Therefore the mould maker learnt how to nick and flatten his wires at any intersections. He had a similar problem with the sewing wires which of necessity had to cross the ones forming the main watermark. We can see that during the fourteenth century these wires are drawn finer, but the mould maker always had to compromise between thin sewing wire that might have a short life and thicker wire that could be detected in the watermarks. Advantage was taken of the fact that the sheet of paper was weaker at the point where a wire made a watermark to deliberately make even thicker wires into tearing wires along which the sheet might be torn into smaller ones or even into envelope shapes. Care had to be taken with these larger wires that the pulp was not caught underneath them so sometimes they might be flanked by ones of much smaller diameter or even flattened at their base.

The art of how the mould maker bent his wires is worth studying on the moulds. A complex crown with crosses, circles and other shapes will have been assembled from very few pieces which double back on themselves in surprising contortions and ingenious ways. On the Continent towards the end of the eighteenth century, thin foil was cut into shapes such as leaves or letters and sewn onto mould covers to give light areas. These were called 'full' watermarks. On larger pieces of foil, drainage holes as well as the sewing holes might have to be drilled to give adequate dewatering.[8] This form of watermark never assumed much popularity in Britain.

In Britain at the end of the eighteenth century and into the nineteenth, we find in laid paper standard types of watermarks in the middle of one half of the sheet. These might be a fleur-de-lys which originated possibly as early as 1285, a fools cap first found in 1479 or a post horn possibly dating from 1670 although this too may go back to the fourteenth century.[9] During the eighteenth century, they all became set in complex shields with crowns on top and the manufacturer's monogram in cursive at the bottom. Another popular mark was the seated figure of Britannia with shield and spear contained within a crowned oval frame, which in Britain replaced the foolscap. These marks indicated the size of the sheet of paper.[10] In the middle of the other half of the sheet would be the name of the manufacturer or stationer who had ordered the paper. It became fashionable to shape these letters in a Roman style.[11]

Below the name might be the date of manufacture. In 1794, an Act[12] was passed allowing no drawback of the duty on the paper of exported books 'unless the Paper shall have visible in the substance thereof a Watermark of the date in the following Figures, 1794, or in like manner some subsequent year'. It was not clear whether this date should be the year of the Act, 1794, or the year of manufacture. Most papermakers followed the second interpretation and regularly changed the dates on their moulds at the beginning of each year. This date should give a fairly reliable guide to the age of nineteenth century English paper.

Wove paper, however, was watermarked more simply. The whole point of producing wove paper was to give a smooth printing surface. One exceptional watermark with the three feathers of the Prince of Wales has been found in the middle of sheets of paper dating from 1811 which was made at the Afon Wen Mill near Denbigh in North Wales. Otherwise the middle of wove paper was left unwatermarked and the papermaker placed his name and date along the bottom edge. The lettering followed the example of laid paper and by 1800 was mostly in a standard form of Roman style.

In 1812, Leger Didot patented a satisfactory way of eliminating shadow zones.[13] Although his specification mentions only laid moulds, his method was quickly applied to wove moulds as well and improved their draining qualities. First, on top of the ribs, he fixed one cover made with the equivalent of laid lines spaced widely apart. Parallel to the ribs, he ran a series of wires over this first cover the whole way across the mould. Finally, on top of both of these, the proper laid or wove cover was sewn on top. The water could flow to the ribs along the lower sets of wires and so Didot achieved a 'due and regular flow or draining off of the water, and uniform disposal of the paper, or stuff, or pulp upon the said mould'.[14] This system of backing wires has been employed ever since on hand moulds.

The next developments in watermarks are associated particularly with the Bank of England. That Bank was created in 1694 when the Government of the day needed money to pay for the war against France. William Paterson, a Scottish merchant, suggested founding a bank which could lend its capital to the Government.[15] £1,200,000 was quickly subscribed but at first there was reluctance to issue notes with the attendant risks. However, as one of the governors wrote in 1697, 'The Custom of giving Notes hath so much prevailed amongst us that the Bank could hardly carry on business without it'[16], the Bank very quickly began to issue notes and has continued to do so ever since. After some notes had been counterfeited in 1695, it was decided to make a mould with a special watermark which, from then onwards, was to be an outstanding feature of all Bank of England notes. It was not until March 1697 that Alexander Merrial was paid for the manufacture of 'paper frames' and it was decided to make paper upon them. From 1697 to 1724, the paper was made by Rice Watkins and Thomas Napper at Sutton Courtney in Berkshire.

During this time, a French refugee, Henri de Portal, had started to make paper in Hampshire at his own mills, first Bare in 1711 and then Laverstoke in 1718. He made quick progress as a manufacturer and in 1724 was awarded the contract for producing the Bank of England paper, which the firm of Portal has held ever since.[17] His paper was a great improvement on that previously made for it was harder, of better texture and the watermark was much clearer than in the old. The Bank's notes, printed with copper plate writing on white paper, relied partly on the distinctive paper but principally on the watermark to prevent people forging them.

This does seem to have been a deterrent to forgers but, in 1758, a Stafford linen draper, Richard William Vaughan, was executed for changing the figures to a higher denomination. In spite of the fact that, in 1773, an Act was passed making the penalty death for copying the watermark, in about 1780, Charles Price, a skilled designer, engraver, papermaker and printer, produced some notes which even the Bank of England itself accepted for a time. The problem that the Bank faced in producing a paper with adequate security may be judged from the fact that between 1812 and 1818, there were circulated 131,331 pieces of forged bank note paper.[18] In the twenty years prior to 1817, there were no fewer than 870 prosecutions connected with bank-note forgery, but 1818 was the culminating point of the crimes for in the first three months of that year, there were 128 prosecutions by the Bank. So the watermark had to be complex but easy to recognise. Then the paper had to be strong to be long lasting which needed a long fibred pulp, but that of course obscured the watermark.

From the late 1790s, the Bank of England endeavoured to find ways of defeating the forgers. One of the people who tried to produce paper that would meet the Bank's security paper requirements was Sir William Congreve. He started experimenting in 1818 and patented his results in 1819.[19] Only some of his ideas were accepted by the Bank.[20] From his patent, we can see that his first objective was to make paper with clear distinguishing marks that would be

> as simple as they can be, so that they may be immediately recognised and understood by the most unlearned persons, that all orders of people may be equally able to distinguish a bad note from a good one, by observing whether the said tests exist or not in any note that may be offered.[21]

First he aimed to improve the quality of the pulp. Instead of cutting the fibres up into short pieces so that the paper became weak and brittle, Congreve tried to mash or rub the rags to pieces to give a long, flaky staple with plenty of wet strength so the paper would be tough and elastic. He also added gum, isinglass, parchment cuttings and the like to obtain tenacity and transparency. In this way, watermarks could be obtained with exceptional clarity that could not be copied by rubbing parts of the paper away, or by trying to impress the watermark after the paper had been made, or by imprinting the watermark with varnish or similar liquids.

The second part of his patent concerned making multi-layered paper. He aimed to introduce a layer of coloured pulp between two outer layers of white paper. His recommendation was that this inner layer should be made from Adrianople red cloth because its colour was clear and permanent as well as difficult to reproduce. These three layers might be formed on the same mould by dipping it in the appropriate vats consecutively and couching off as one sheet. But he really recommended that three moulds should be used which enabled any one of them to be watermarked or plain. Also it would then be possible to make the mould on which the middle layer was formed of a different shape or size from the others, and it could be watermarked as well. If separate moulds were used, then they had to be couched with a special frame in which the moulds were fitted so that the layers of paper were imposed upon each other in exactly the same position for each sheet of paper. The consistency of the stuff in the vat was to be as thin as possible which might mean that a double dipping of the mould was necessary to develop sufficient depth of pulp to form the darker parts of the design. Although made as multi-layered paper, it would still be thin and strong. The coloured pulp forming the middle layer would be contained within the heart of the sheet, brilliant, distinct, but permanent and indestructible. This coloured watermark would be seen only when the paper was held up to the light.

Portals, papermakers to the Bank of England, made eight sheets of this paper in ten minutes but that was by using only one mould. To have made the paper on three moulds with three couchings proved to be too expensive, even when eight notes were formed on the mould at once. So the Bank did not pursue Congreve's suggestions for multi-layer paper. However, at Silkeborg in Denmark, multi-layered

The Art of Watermarking

7. The deckle has been lifted off this sample mould with the watermark embossed
in a wove cover.

paper was made by hand from two vats with different types of pulp in them. At
Tumba in Sweden, one mould was dipped first into a yellow pulp and placed
under a machine which scattered coloured fibres over part of it. It was dipped in
a white pulp and couched, which left the yellow side uppermost. On top of that,
there was couched a second mould which had been dipped in the white pulp. This
gave a paper with a white surface on either side, yellow in the middle and a band of
security fibres in one part too. So some of Congreve's suggestions for multi-layered
paper were taken up later.

Some of Congreve's ideas for improving the quality of banknote paper were
adopted by the Bank of England so that its paper became distinctive. It was described
in 1857 in the following way.

The paper of the notes of the Bank of England was distinguished by its colour – a peculiar white, such as was neither sold in the shops nor used for any other purpose; by its thinness and transparency, qualities which prevented any of the printed part of the note from being washed out by turpentine, or removed by the knife, without making a hole in the place thus practised on; by its characteristic feel, a peculiar crispness, by which those accustomed to handle it distinguish the true notes instantly; the wire or watermark, which was produced in the paper when in the state of pulp and which was easily distinguished from a mark stamped on after the paper was completed; the deckle edges – the mould contained two notes placed lengthwise, which were separated by a knife at a future state of the process – this deckle producing the peculiar effect seen at the edges of uncut paper, and this edging being caused when the paper was in a state of pulp, precluded any successful imitation after the paper was made; also by the strength of the paper, which was made from new cotton and linen. In its waterleaf or unsized condition, a bank note would sustain 36 pounds; and when one grain of size had been diffused through it, it would lift 100 pounds.[22]

The first watermarks in the Bank of England notes had a border with a loop pattern running round the edges of the sheet. In 1724, the words 'Bank of England' were added just above the bottom border. In 1798, William Brewer, a watermark mould-maker of Maidstone, approached the Bank with a plan for making a change in the design of the watermark. He suggested waved lines. During the trials and experiments, a further protective device was introduced by including in the design the denomination of the notes.[23] The experiment was successful and, in September 1801, the following advertisement appeared,

All the one and two pound notes issued by the Bank of England, on and after the first of August will, to prevent forgeries, be printed on a peculiar and purposely constructed paper; consequently those dated 31st. July, or any subsequent day, will be impressed upon paper manufactured with waved or curved lines.[24]

A zig-zag pattern of laid lines could be made by taking alternate chain lines and pulling them in opposite directions so that the laid lines bent at the chain lines into a series of 'W's. To produce the curved laid lines in the Bank of England notes, the laid wires had to be bent over a tube or rod so they formed semi-circles between each chain line which gave a sort of ripple effect in the watermark. The manufacture of watermarks with similar patterns was forbidden. By assembling a background of sections of these different effects, and then including the words 'Bank of England' twice and 'Five', a very complex watermark was achieved and so extremely difficult to forge. However, to produce and maintain such a watermark was a mould maker's nightmare because the stitching binding all these small wires to the cover was liable

40

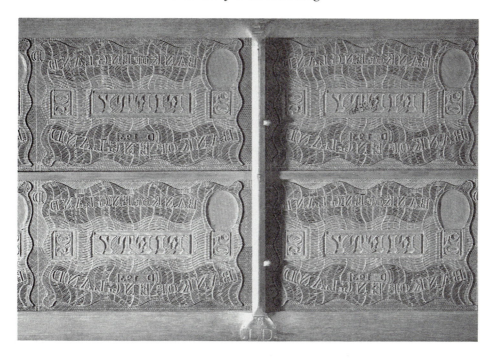

8. Brewer and Smith's patent watermark pressed from a sheet of metal for a
Bank of England note.

to break allowing the pieces to fall off. Also it was difficult to duplicate the wire profile exactly in each mould and each mould made four pairs of sheets at once and so required eight wire profiles. In a pair of watermarks for five pound notes, there were 8 curved borders, 16 figures, 168 large waves and 240 letters, all of which had to be separately secured by the finest wire to the waved surface. There were 1,056 wires and 67,584 twists for the wire profile itself and about the same number for the backing layers, so each mould had hundreds of thousands of stitches.[25] When in use, each mould was checked three or four times a day to see that the watermark remained in order.

The problem of making these complex watermarks for the Bank of England was solved by the patent of Brewer and Smith in 1849.[26] Basically, their method was to make a steel die and press out the wire profile from it. Unfortunately their patent does not mention how the die was engraved. Today, the practice is to first draw out the watermark design on paper, making allowances for distortions to compensate either for the unequal contraction of the paper as it dries or on a paper machine for the dandy roll dragging and so elongating the watermark. For example, a circle in the finished paper may have to appear as an elipse in the design. Then the design is curved out of a block of wax in which the light and dark areas of the finished watermark are changed into high and low parts of the wax. This requires considerable skill, for in producing the required image in relief, it needs the sculptor's scapel to remove only a fraction too much wax for the design to be ruined, necessitating the work having to start all over again. The wax is coated with a conducting material such as graphite so that copper can be deposited upon it by electroplating. The wax is melted, leaving a copper master in low relief which is backed by filling it with lead. From this master, the steel dies can be prepared by electric etching techniques.

Brewer and Smith merely say,

> The design is cut and engraved upon the surface of a steel die in the ordinary manner, care being taken in cutting the die that the parts of the design which are to be the lightest in the watermark are to be cut deepest upon the die, the parts of the design which are to be darkest in the watermark, or nearest to the general substance of the paper, being cut to the least depths on the die.[27]

After the die had been hardened and cleaned, Brewer and Smith fitted it on the hammer head of a steam hammer, today replaced by hydraulic presses. To form the obverse of the die, they forced or stamped the die itself into a red hot copper plate or tin in a liquid state. Repeated blows might be necessary to obtain a full impression. Having formed their matching halves of the die, they could press out the profile for the watermark itself from either a sheet of copper or bronze metal or today a sheet of woven phospher bronze wire cloth.

At first, pressing out profiles from a sheet of metal assumed more importance because that was how the complex watermarks for the Bank of England five pound and other notes were made until production by hand ceased in the 1950s. As they came off the press, these stamped sheets had no drainage holes so they were clamped in a special holder to have the backs filed off. Other parts might be cut away by rotary cutters. Instead of a grid of backing wires to assist with even drainage, a second plate would be stamped out from another die engraved with the equivalent of the lower grid of wires. This also had to be cut and filed to give drainage holes and then the pair could be clamped together by screws.

Brewer and Smith claimed that, with their method, a large number of identical wire profiles could be produced, which was impossible by bending and sewing wires. Their watermarks could contain facsimiles of signatures, seals and other devices hitherto unobtainable. Their moulds lasted longer because there was no stitching and could be cleaned more thoroughly because they were stronger. Also, because there were no stitching wires to block the holes, drainage was better, so the quality of the paper was improved, being more even in texture and with fewer spots. The advantages of such long lasting and identical moulds can be judged from the fact that in 1855, the Bank of England was issuing nine millions of notes per annum, representing nearly three hundred millions of pounds.[28]

Recognition of Brewer and Smith's work was given in 1851 when the Bank sealed an agreement acquiring the sole right to their invention. The first notes made on the new moulds were issued in 1855.[29] At Threadneedle Street, the Bank of England has preserved a handmould with its set of eight watermarks. It is a piece of exquisite craftsmanship, in the tradition of the highest quality of British mould makers. A projecting lug compels the vatman to fit the deckle in one way only. Nearly all the wood is sheathed in brass to prevent wear, but this makes the mould very heavy. The wire profiles are still in excellent condition.

In the longer term, the other method of pressing out a woven wire cloth for making wire profiles has proved to be more important. Using wove wire as part of a watermark passed through two stages. First a small oval or other shape might be let into the surface of a laid cover. This may have been introduced in France around 1793 in the ill-fated Assignats.[30] Where this was lower than the rest of the surface, it gave a 'positive' or darker watermark. Certainly by 1812, Johannot in his French mill had placed a wove background behind some of his portraits made from wire.[31] It was possible that at Fabriano wove wire was first shaped into a three-dimensional watermark soon after 1800. Here the required relief was carved out of a piece of fine grained wood such as box. Then annealled wove wire was laid across it and pressed into it by means of a wooden or bone pin so that it would closely follow the shape of the wood.[32] Brewer and Smith's contribution may have been pressing out the wire between dies. Wove wire can be shaped to give graduations in level so that the thickness of the paper can vary smoothly and not suddenly. This produces what have been described as intaglio or three-dimensional watermarks. The Queen's portrait in the Bank of England's present notes is formed by this process.

Over the years, the three-dimensional watermark has developed into an art form in its own right. In 1867, T.H. Saunders of Dartford displayed paper with three-dimensional pictures which produced an effect almost incredible to those who knew only the earlier type of wire profile.[33] Designs may include both three-dimensional marks and ones made from wire to give quite remarkable patterns. Portraits of famous people have been especially popular and famous pictures have been reproduced too. To achieve the best results, the fibres need to be short and the pulp tinted to enhance the contrast. In handmade paper, the mould may be dipped two or three times to fill the deeper places. The design has to be carefully prepared because, as the wire cloth is pressed, it stretches and the holes are enlarged. To achieve the full range of height and depth, the cloth may have to be annealled between pressings. To give sufficient depth for the darker parts, the mould framing may have to be cut away, while in the lighter parts, the cover may need additional support. In a portrait, the highlight on a person's nose will be a weak part in the sheet which may break in couching or drying. Such watermarks need careful planning in their original construction to see that the drainage will function and that the sheet can be released cleanly from the mould. Great care must be taken in forming the sheet, couching and drying it afterwards. These comments apply equally to watermarked paper made on machines but the results can be very beautiful. By 1869, William Frederick de la Rue had adapted the three-dimensional watermark for application to dandy rolls[34] and cylinder mould machines followed later.

Where a large number of identical watermarks made from wire profiles is needed today, for say the cover of a dandy roll or the cylinder in a mould machine, the galvano or electrotyping process is used. This method had been introduced certainly before 1855.[35] After the design has been agreed, each detail is formed by hand in copper wire and the individual sections are pinned to the original working sketch. Then the various parts are wired together and the whole is soldered onto a copper plate which becomes the 'master' die for pressing into blocks of wax. The wax is dusted heavily with fine powdered copper. Removal of the surplus metallic dust from the surface of the wax leaves the design still covered with powder in the impressed areas. The wax can then be plated with copper in an electroplating bath which takes about twelve hours. If the watermark has many separate parts, such as the letters of a name, they can be joined together with bars so that just one single electrotype is produced. The wax is melted away, leaving the electrotype. Hollows in the back may be strengthened by filling with tin or solder. At first, electrotypes, were sewn onto the covers but by 1870 T.A. and C.D. Marshall, descendants of John, began soldering them onto dandy rolls.[36] The linking bars are of course cut away, leaving just the actual wire profile. Such watermarks are more firmly attached, are practically identical, but somehow lack the individuality of earlier ones bent from wire.

IV

A NEW INDUSTRY
EMERGES, 1500–1800

Rags make paper,
Paper makes money,
Money makes banks,
Banks make loans,
Loans make beggars,
Beggars make rags.

How true this was to prove of the early paper industry, yet during the years 1500 to 1800, it was to grow in Britain from nothing to a position where mills had spread across most of the country. These three hundred years saw an enormous expansion in economic and industrial activity in so many spheres which created ever increasing demands for paper. Not only did the population increase but wealth generally increased too. This was partly through a growing overseas trade linked with the discovery of the New World and trade routes to the east but also at the end of this period to the beginning of the Industrial Revolution which would dramatically change the paper industry itself. Not only did British production supplant imports of all types, but it met the increasing demand and also started to export some grades of paper.

It was only in 1492 that Christopher Columbus sailed west to discover what was in fact part of America. England was late in joining the voyages of discovery but Francis Drake circumnavigated the world in 1586. More important was the establishment of merchant trading companies with, for example, the Muscovy Company around 1553, the Levant Company in 1581 and the famous East India Company in 1600.[1] The Pilgrim Fathers left Plymouth in 1620 to found what has been considered the first of the colonies in North America, an area that was to be so important in later development of trade and industry. Expanding commerce needed more paper, and in particular good quality white paper; paper on which to write orders, paper on which to issue bills, paper on which to keep accounts, record transactions, stocks and so on. Then there was an increasing use of paper of lesser quality for wrapping goods to keep them clean or in which they could be contained. Also new types of paper to meet new needs were developed so the paper industry presented a picture of expansion, change and challenge.

We can see also an expanding demand for paper caused by improvements in other spheres. Improved communications through better roads, ships and harbours stimulated the writing of letters and so the demand for paper. Travellers had always carried letters for their friends but governments needed their own courier services. It was out of these that in 1635 the State service in England was extended by Royal proclamation to cater for the conveyance of public correspondence upon payment of fixed postage rates. A penny post was established within London in 1680. At first, the postal system mainly linked London and towns in the rest of the country, but in 1720 Ralph Allen, of Bath, was granted a monopoly of the posts between provincial towns and established a system of postboys riding on horseback. This did not work too well because they were at the mercy of highwaymen who abounded on the roads during the eighteenth century. Then came the development of turnpikes and the introduction of stage coaches which competed for carrying mail. In 1784, John Palmer, also of Bath, persuaded the Government to accept the idea of a Mail Coach service protected by armed guards. From 1787, the mails were carried in a new design of coach constructed by John Bezant which increased reliability and regularity.[2] In fact, when reading the correspondence between the headquarters of the firm of Boulton & Watt in Birmingham and its engine erectors scattered all over the country, it is possible to see that answers could be sent within a couple of days of the date of the original letter, something rarely achieved with our postal services today. So by the end of the eighteenth century there was a good postal service throughout most of the country.

Paper on which letters and correspondence was written was not any great size and so presented no special production problems in the days of the hand industry. Neither did that used in books because it was limited by the capacity of the printing presses. In letterpress printing there was little technical improvement from the presses of Caxton's day, but there was an enormous increase in numbers with most large towns eventually having their own printers. When this type of printing was being developed in the middle years of the fifteenth century in Germany, the screw press for pressing cloth or the wet sheets of paper was adapted but differed in three respects. The first was in the speed of operation. On a hand printing press with three people working round it, the speed of making the impressions was increased from around 100 to 250 impressions an hour. This has to be compared with the papermaking press which might be used two or three times a day and might be left under pressure for a considerable period to squeeze out the water. Then the screw of the printing press did not have to be turned round and round but had to be quick-acting, so it had a thread with two or three leads and put the pressure on in a quarter of a turn. During the seventeenth century, lever mechanisms were invented to replace the screw so the head of the press descended vertically and would not twist and smudge the print.

The final difference was the need for absolute flatness and parallelism between the bed and the platen of the printing press. In papermaking and textiles, this was not vital, but on a printing press, parts of the type would not be printed if the surfaces were not parallel. Stone beds and wooden platens were common by 1700 when the average size of the platen was 18 x 12 in. (46 x 30 cm.) or 216 sq. in. Later, the largest wooden press took a size 26 x 20 in. (66 x 51 cm.) or 520 sq. in. The strength of the impression mechanism and the platen itself were the limitations on making bigger presses. If larger areas of types had to be printed, one solution was to extend the bed to take two sets of type, print first one half and then move the bed on to print the second.[3] In 1772, William Haas of Basel built a press with stone base, cast iron bed, cast-iron platen and parts of the framing of cast iron with a printing surface of 340 sq. in. It was not used much because the framing was weak and cracked. Therefore the size of the sheets of paper needed for this method of printing right up to 1800 did not exceed what could be made on a mould handled by one vatman.

Growing wealth and changes in the way of life were increasing the number of people who both could read and who also had time to read. By the early eighteenth century, it had become the mark of middle-class women not to work and they would have been intolerably bored without the creation of the novel. There was also the movement towards free expression which was to make eighteenth century Britain widely envied. This was to give rise to the newspaper, and the first one to be published daily in London was the *Daily Courant* in 1702. An estimate of 1704 lists nine newspapers throughout England with a total circulation of 45,000 per week. There were weeklies or bi-weeklies in Bristol (1702), Exeter (1707), Worcester (1709), Nottingham (1710) and of course the famous *Spectator* which Joseph Addison began to publish in London in 1711. During the 1730s, circulation of one edition might reach over 10,000 copies.[4] So newspapers became, and still remain, important customers for paper.

Although they did not use anything like the volume of paper when compared with newspapers, pictorial illustrations grew in importance and popularity during the eighteenth century both within books and also as wall decorations. Printing with patterns carved on wood-blocks existed long before printing with movable type but this particular method of reproducing illustrations suffered an eclipse during the early years of the eighteenth century until it was revived towards the end of this period by Thomas Bewick of Newcastle upon Tyne. In 1662, the mezzotint was first used in an English book and the art of engraving on copper plates had been mastered by the end of the seventeenth century. Throughout the eighteenth century engravers used both the line or engraving and the etching techniques, or a combination of the two.[5] One of the famous illustrators was William Hogarth, who achieved greater success and fame as an engraver than as a painter. Much of his best work was engraved before 1750, including 'The Rake's Progress' in 1735. In fact his illustrations were so popular that in 1735 he had to seek an Act of Parliament to protect the commercial potential of his prints and prevent piracy.[6]

Books illustrating natural history, travel, architecture and similar subjects with copper-plate engravings became popular, especially towards the end of the eighteenth century. During this period, learning how to draw was considered part of a good education, especially for young ladies. To meet the demands of a leisured society, Rudolph Ackermann opened a drawing academy at 96 Strand in 1795 where he also began publishing and selling large illustrated books, prints and later periodicals. In 1798, he renamed his firm the 'Repository of Arts' and sold drawing supplies as well as fancy paper goods. This popularization of the arts was reflected in the interest in painting, particularly with watercolours. The Society of Painters in Water-Colours was formed in 1804[7] and of course stimulated the demand for artist's papers.

Of much greater importance was commercial application of copper-plate engraving in the reproduction of maps, navigation charts and even technical drawings. No ship could navigate without its charts and their range had to expand as the voyages of discovery opened up the world. Through the work of John Harrison with his marine chronometer for example, improved navigational instruments became available which necessitated resurveying many coastlines, a work in which Captain James Cook was involved before his death in 1779. Equally, if more than a very few copies were needed of an architectural or engineering drawing, the quickest way to reproduce them was with the copper-plate engraving. The type of press on which these engravings were printed will be discussed later because this was one of the cases in which sheets of paper larger than could be made by hand were employed.

The increasing demand for banknote paper has been discussed already (see p. 37f). This was important for the development of high quality security paper. A trade important for demand on a large scale was the wallpaper industry which will be examined separately. There remains the use of cheaper grades of paper particularly associated with industry. One of the influences on the distribution of papermills within England was the localisation of certain types of paper manufacture in industrial areas where there was a growing demand for paper. Birmingham, and to a lesser extent Wolverhampton, became noted for papier mâché and japanned work for which the basic paper materials probably came from mills in the west Midlands. Then several woollen manufacturing districts such as Exeter and the West Riding of Yorkshire provided an important market for pressing, finishing, packing and wrapping papers and boards. The cloth was pressed with boards inserted between the layers of fabric and then despatched with paper packing to protect it. The principal cotton manufacturing districts in Lancashire and Derbyshire also stimulated the manufacture of paper both because of the availability of raw materials for papermaking from the linen and cotton waste and also as a market for packing papers.[8]

Then there was a number of mills engaged in the manufacture of card and board. Playing cards were introduced from an early date in the fashionable watering places such as Bath and these needed board of high quality made from sheets of paper glued together. Lesser quality pasteboard began to be used in bookbinding, replacing wooden boards during the sixteenth century. This again was made by gluing together sheets of paper. Boards were being produced in England by 1700 at least. A 'Brown Board Mill' had been established at Bourne End in Buckinghamshire before 1719 and by the end of the eighteenth century there were several board makers in this vicinity because it was convenient for the London market.[9] There was also millboard originating at about the same time. This was made from the same fibres as pasteboard but was formed in a much thicker layer on the mould in a single sheet and then milled or rolled under pressure.

For wrapping up powdered grocery articles like sugar, paper once again was in demand because there was nothing else from which to make suitable small containers. Traditionally sugar was always wrapped in blue paper and the earliest patent concerning paper was taken out by Charles Hildeyerd on 16 February 1665 for 'The way and art of making blew paper used by sugar-bakers and others'.[10] Then in 1691 Nathaniel Gifford patented 'A new, better and cheaper way of making all sorts of blew, purple and other coloured paper'[11] but unfortunately in neither case have any details survived.

In spite of this growth in different types of paper, its consumption in Tudor and early Stuart times never made it an article of common everyday use. A quire (24 sheets) of white writing paper cost about 4 d. or 5 d. which was about as much as a labourer's daily wage and so it is unlikely that much of it could have been purchased by any save the rich or those concerned with administration and trade.[12] An extremely rough estimate of annual consumption per head of white paper, at the turn of the sixteenth century, suggests a figure of about ¼ lb., and a rather more accurate one for the five years 1714–18, covering all types of paper, giving 1½ lb. per head per annum. Although this had risen to 2½ lb. by the end of the century, paper was still far from being a common item for most people. By modern standards, these proportions are minute for in 1955, consumption per head of all types of paper in Britain was nearly 130 lb.[13] Even as late as 1700, the greatest import of paper and the consumption of white paper was in the capital, London, where between 1699–1703, London's share of imports was 97 per cent, with all the other ports in England and Wales taking the rest.[14]

In addition to rising demand through a steady increase in consumption per head, the population of the country also grew steadily, further stimulating production. Around 1500, the population of England and Wales may have been 3,000,000 and in 1600 possibly 4,500,000. It is a fair estimate that in 1700 the figure was between 5,000,000 and 5,500,000 and in 1750, about 6,500,000. Between 1750 and 1801, the year of the first census, there was a growth of 40 per cent to between 9,000,000 and 10,000,000 people.[15]

Rising demand was an incentive to establish an indigenous paper industry but it had to make its way in competition with other industries needing water and waterpower to drive their machinery. The traditional types of waterpowered industry had been well-established in England for centuries. From the Domesday survey compiled before 1086, it has been deduced that there were 5,624 watermills in 3,000 different locations in those areas actually surveyed.[16] John Tate had converted a mill, presumably a corn mill, which had been mentioned in Domesday. Sir John Spilman in 1588 took over two mills known as 'The Wheat Mill' and 'The Malt Mill'.[17] Advantage was taken of the decay of the woollen industry in Kent to convert fulling mills into papermills. One such example was the famous Turkey Mill near Maidstone adapted by George Gill shortly after 1670.[18] Sometimes paper mills were run in conjunction with other industries when the water resources of a stream or river allowed. On the river Derwent in Derbyshire, during the eighteenth century, papermaking featured among other industries at the sites of both Masson and Darley Abbey mills where the later cotton textile industry drove out the paper.[19] Growing industrial expansion increased the competition for good waterpower sites during the eighteenth century and this was barely alleviated by improvements to waterwheel design instituted by people like John Smeaton during this same period.[20]

Because the next mills established in England after John Tate's produced brown paper, it is difficult to discover both when they started or when they failed and therefore to know how continuously paper has been made in this country. There is no evidence for another paper mill until Thomas Thirlby, Bishop of Ely, tried one at Fen Ditton near Cambridge, possibly from 1554 to 1559. There was another one at Bemerton near Salisbury begun in 1554 or 1569 and a third at Osterley, Middlesex, started by Sir Thomas Gresham in about 1574. Bemerton was the only one of these mills to survive for more than a few years[21] for the one at Osterley had failed by 1593. Richard Tottyl, a London stationer, petitioned in 1585 for a privilege of the sole right of making white paper in England and for the prohibition of the export of rags. He stated that he and other members of the Stationers' Company had tried to set up a mill twelve years earlier and complained that French papermakers bought up the supplies of English rags. Neither of these attempts succeeded.

1588 saw not only the defeat of the Armada but also significant advances in the development of the paper industry in both England and Scotland for good quality paper may have been produced in England continuously since then. It was left to John Spilman, a German who became jeweller to Queen Elizabeth I and then King James I, to establish the first mill for making white paper that approached anything like success. In 1588 he was granted a lease of two royal mills on the River Darenth near Dartford and converted them for making paper. The work-force he imported from Germany. In the following year, he was granted monopoly powers for buying or dealing in rags and for licensing others to make paper, including those already making brown paper. In 1597, he was granted another privilege for a further fourteen years. Possibly his chief claim to fame lies in the doggerel verses written by Thomas Churchyard in 1588.

If paper be, so precious and so pure,
   so fitte for man, and serves so many wayes,
So good for use, and wil so well endure,
   so rare a thing, and is so much in prayes:
Then he that made, for us a paper mill,
   is worthy well of love and worldes good will.
And though his name, be Spillman by degree,
   yet *Help-man* nowe, he shall be calde by mee.[22]

Churchyard's claim that Spilman employed a workforce of 600 is surely an exaggeration when later most one vat mills would have only about ten. The mill was visited by James I in 1605 when Spilman received a knighthood. It remained at work probably making mostly brown paper until the death of his son John in 1641[23] by which time there were others established in England.

Looking at what happened north of the border in Scotland, we find that in 1588 there was a growing realisation that Scotland would be a profitable country in which to make paper. In paper, as well as in metallurgy and mining, the technically advanced Germans by the late sixteenth century were seeking opportunities abroad where they could exploit their knowledge, and the favourable attitude of James IV towards industrial development offered an incentive for such people to settle in Scotland. In 1588, James IV granted a 'liberty' to Peter Groot Haere and others 'to set up this art of making paper of all sorts within this realm' for a period of nine years.[24] This liberty does not seem to have been taken up but on 5 December 1590, the Register of the Privy Seal states that a monopoly was granted to 'Pietter Gryther [Peter Groot Haere] and Michaell Keysar, Almanis, paper makeris' appointing them papermakers to the King for nineteen years.[25]

By 1590, papermaking was being carried on at the Wester Mill of Dalry on the Water of Leith near Edinburgh. By a contract of 3 May 1594, the Russells, owners of Dalry Mills, agreed to provide further accommodation by raising the mill walls by eight feet and installing a drying loft.[26] By 1595, Keysar was in partnership with another German, John Seillar, and they agreed to instruct apprentices chosen by Gideon Russell, but they were not to give any assistance in building other paper mills. The Dalry Mill was mentioned in 1605 but nothing is heard of it again until 1673 when it was leased by six Edinburgh merchant burgesses who were manufacturing paper by 1675 when French craftsmen were introduced.[27]

In all probability, the original Dalry Mill failed around 1605 and no paper was made in Scotland until a second mill was started, probably in 1652. It was situated at Canon Mills, also on the Water of Leith. According to an Edinburgh council minute, John Paterson was the operator and tackesman in 1659 when a visit was to be paid to the mill by the Dean of Guild and others and they were to report to the Council.[28] It is probably from this time that paper has been made continuously in Scotland.

Meanwhile, in England, it would seem that forty-one paper mills existed between 1601 and 1650. Twenty-three of these were within thirty miles of London and the others scattered widely across the country. By 1675, a further twenty-two mills had been added to the list and the number had risen to 116 mills or sites by 1700.[29] In 1712, the Excise authorities gave the totals of papermakers in England and Wales as 209, and seven in Scotland.[30] Although mills were numbered by the Excise authorities, there is difficulty in confirming that they were all actually in operation, so the following figures are approximate only. In 1738 there were not more than 278 mills with 338 vats.

YEAR	LICENCES
1775	345
1785	381
1790	389
1795	407
1800	417[31]

The establishment of this industry was achieved against fierce foreign competition. Imports of brown paper reached a peak in the early decades of the seventeenth century and dwindled to a trickle by the 1670s when home production had captured the markets. In the import of white paper, there was a marked rise until the later years of that century and thereafter a decline. In the later decades of the fifteenth century, France was ousting Italy as the main source of supply for white paper, a position which continued for the first three-quarters of the succeeding century. However by that time, the Netherlands had begun to take over. The Dutch at first acted as middlemen, for they financed the production of paper particularly in France and then shipped it to England. During the seventeenth century a paper industry was developed in Holland beginning in 1586 and, through their use of the roll beater or Hollander, the Dutch began to overtake the French, so that during the eighteenth century Holland was England's main supplier with its own production.[32] Gradually, as the quality of English paper improved, the Dutch in their turn were ousted so that, in 1720, it seems likely that England was producing about two-thirds of its home consumption, and by 1800 had become not only self-sufficient but had entered the export markets.

It would seem that manufacture of white paper had come to a halt in England around 1641. One problem may have been the lack of a linen industry which could provide suitable raw materials and doubtless the Civil War disrupted papermaking. Later various people tried to enter white paper production, and two took out patents. The first was Eustace Burneby in 1675 who was going to make 'all sorts of white paper for the use of writing and printing being a new manufacture never practised in any our kingdomes or dominions' (he had not heard of Tate or Spilman), and the other was John Briscoe who, ten years later, would produce

English paper for writing, printing, and other uses, both as good and service-
able in all respects, and especially as white as any French or Dutch paper.[33]

In 1686, a patent was granted to the Company of White Paper Makers in
England which comprised both Frenchmen and Englishmen. In 1690, its charter
was confirmed and prolonged by Act of Parliament and it was given a monopoly
of this production. The company probably had five mills in 1690 and eight by
1698[34] and seems at first to have done quite well. However, other manufacturers
protested that there were above a hundred mills in the kingdom the majority of
which had been employed for the greatest part of their time making white printing
paper and several had been making white writing paper as well. While this was no
doubt an exaggeration, it does show that some white paper of quite good quality
was being made in England well before the end of the seventeenth century. The
manufacture of banknote paper starting in 1697 might be cited as an example (see
p. 37). North of the border, the Society of the White-Writing and Printing Paper
Manufactory of Scotland was floated as a joint-stock company in 1694. Mills were
established in 1696 at Yester, in East Lothian, and at Braid, near Edinburgh,
where good white paper was being produced which continued to work well into
the eighteenth century.[35]

In the half century between roughly 1670 and 1720, the situation changed. A
major expansion in English paper production took place and the industry took firm
hold in Scotland and Ireland too. A calculation made in 1696–7 suggests that the
native industry was producing about 42 per cent by value of total consumption. By
the turn of the century, the total production for England and Wales was hardly more
than 2,500 tons with imports of about 1,000 tons. By 1718, production was about
300,000 reams, or about 71 per cent by quantity from a total consumption of some
418,000 reams.[36] Of this, well over half comprised brown paper in the earlier years
but this proportion declined somewhat by 1800. By 1782, out of a total production of
900,000 reams, about 480,000 were brown and white-brown. In 1800, these figures
were total production 1,200,000 with 570,000 reams of brown.[37]

A statement of 1712 suggests an average of about 1.2 vats per mill and the
figures of the Excise authorities in 1738 support this. This would link up with
the type of beating equipment available then, where a waterwheel driving a set
of stampers would provide enough stuff for one vat. The work-force employed
directly in the industry around 1700 would not have been more than 2,500 to
3,000.[38] Between 1738 and 1800, paper output rose nearly fourfold but the number
of mills only doubled. This was probably due to the higher productivity achieved
by the introduction of Hollander beaters. In 1805, it was estimated that there were
762 vats when 461 paper maker's licences were issued, giving an average of 1.5 vats
per mill. The great majority of mills, possibly as many as three-quarters, always had
only one vat. Of those which had two or more, most were situated within 60 miles
of London. By 1800, there were 4-vat mills at West Mills at Newbury, Berkshire,

Padsole in Kent, Throstle Nest near Manchester, Wilmington in the East Riding of Yorkshire and Wolvercote Mill near Oxford. Only Turkey and Upper Tovil Mills in Kent were equipped with 5 vats.[39]

Technological developments during this period will be examined in the order in which paper is actually made, starting with the raw materials and ending with finishing. It is thought that the early papermakers obtained most of their rags from within the country. In the 1790s, Crieff paper mill in Scotland consumed over two hundred hundredweight of rags per annum.[40] But as the industry expanded, so rags had to be imported from overseas. The first Customs ledgers starting in 1725 show that the Low Countries and France were important sources of supply to begin with. Then after 1750, the majority came from Germany but, by the end of the century, rags were arriving from America, Eastern Europe, Scandinavia and Russia. This may have been caused by the Napoleonic Wars but may also reflect the growing scarcity of rags which was making itself apparent by that time.

Import of Rags, 1725 – 1800

YEAR	TOTAL IN TONS
1725 – 30	192.6
1731 – 35	230.0
1736 – 40	695.4
1741 – 45	422.8
1746 – 50	808.6
1751 – 55	1,142.6
1756 – 60	1,267.0
1761 – 65	1,868.2
1766 – 70	2,686.0
1771 – 75	3,289.2
1776 – 80	2,805.6
1781 – 85	3,203.0
1786 – 90	4,729.4
1791 – 95	4,598.0
1796 – 1800	3,404.6[41]

The supply was partly alleviated by the introduction of the cotton fibre to papermaking. Flax was preferred because its fibre was longer and stronger and the fibre wall straighter and thicker. When beaten during the papermaking process, the fibre would splinter rather easily along its length, bursting out into small fibrils. These fibrils would interlock and impart additional strength when the paper was formed. Cotton, by contrast, had a thinner fibre wall, which permitted the fibre to

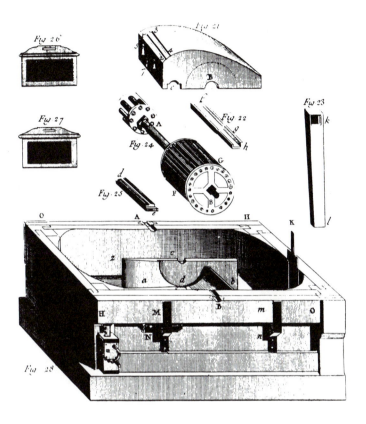

9. The parts of a Hollander beater have been separated in this drawing from Diderot's *Encyclopédie*, 1779. Above the trough on the left are the washing screens which fit in the hood at the top. The roll in the middle would be lowered onto the breast below.

collapse, twist and become ribbon-like. While it did not have the easy fibrillation of flax, and hence the delay in its adoption, its twist produced bulk and opacity as well as softness. With the dramatic expansion of the cotton textile industry through the spinning inventions of Hargreaves with his jenny, Arkwright with his waterframe and Crompton with his mule which enormously increased the supplies of yarn after 1780, there was much more cotton waste from the mills as well as cotton cloth in the country.

Although the introduction of chemicals for bleaching occurred at the very end of this period, this really belongs to a subsequent phase of the industry and will be examined later. In the early industry, the quality and colour of the paper depended entirely upon selection of appropriate raw materials. Rags were usually sorted, graded and cut up by women. This was a dirty and often dangerous job because they could be contaminated by infectious diseases. To achieve the correct grade of paper, not only did the correct mix of different types of fibres have to be determined but the spinning and weaving techniques in different fabrics could alter the character of the paper too. Most mills had their own recipes for blending different types of rags which were kept a close secret.

To overcome the shortage of rags, other substances were tried. A vein of asbestos was found in Anglesey in 1684 and Edward Lloyd suggested that paper might be made from it. In 1719, Réaumur presented a treatise to the French Royal Academy in which he observed that wasps used wood filaments to make their nests which resembled paper and that these insects could instruct us how to make paper from wood.[42] Still on the Continent, J.C. Schäffer published a six-volume work in 1765 containing samples of paper made from a wide variety of plant fibres such as mosses, vines, thistles and much else. Meanwhile in England, in 1728 Mr. Derham showed to the Royal Society very good brown and white-brown paper made from nettles and other weeds. Then in 1788, at Millbank Mill in Cheshire, strong paper was made from the bark of the sallow for which the papermaker, Thomas Greaves, was awarded a silver medal by the Society of Arts. At the end of the eighteenth century, this Society offered a series of awards for paper made from vegetable substances but there were no practical results.[43]

Probably the most important invention during these three centuries was the Hollander beater or 'engine' in papermaking terms. As its name implies, its country of origin was Holland and it appeared during the middle of the seventeenth century, but nobody knows who developed it or precisely when. It certainly had been invented before 1673 when a request made to the Dutch government for a patent was refused. After this it spread quickly in the region north of Amsterdam because, when the iron bars were replaced by bronze, it enabled the industry there to make white paper. It consisted of an oblong wooden tub with a dividing wall down part of the centre so the rags or pulp could circulate. The rags were beaten by bars fixed parallel to the driving axle into the circumference of a solid wood drum or roll. These beat the rags against another set of bars fixed in a bedplate in the bottom of the trough. From here the rags were lifted over a breast to be properly mixed and circulated. The rags were churned round and round until they became pulped. The roll had to be covered with a hood because, while the pulp passed over the breast and fell back into the trough, the speed of the roll was sufficient to fling some water back over the top where it could pass through wire filters and run out of the Hollander. Clean water was poured into the Hollander to wash the rags and was removed through these screens in the hood together with the dirt.

Hollanders cut and lacerated the fibres, producing a different quality of pulp from stampers which rubbed and frayed the material leaving longer fibres. Speed was the great advantage of the Hollander and that is what the papermarkers in the Zaan region of Holland with their windmills needed in order to be able to compete successfully with the abundant waterpower of the French and Germans. In 1725,

Keferstein, a papermaker in Saxony wrote, 'The Hollander in Freiberg furnishes in one day as much as eight stamper-holes do in eight days'.[44] By the end of the century, it was stated that it took forty of the old stampers twenty-four hours to reduce one hundredweight of rags to pulp. In the same time, the Hollander could prepare twelve times this amount. The reason for this was that the roll of the Hollander, with 20 or 24 bars around its circumference and revolving at 120 r.p.m., would make 180,000 cuts per minute.[45]

The introduction of Hollanders enabled mills to increase in size because one windmill or waterwheel could be coupled more easily to a range of Hollanders than to rows of stampers. Also Hollanders could be built to take a greater capacity. Another Dutch invention which appeared at about the same time was the power-driven pestle or 'Kapperij' with knives on lower ends of vertical shafts which cut rags in a rotating tub.[46] This again helped to increase the rate of production. While both the Hollander and the rag cutter were evolved first in windmills, soon they were adapted for watermills and then spread to other papermaking areas. An

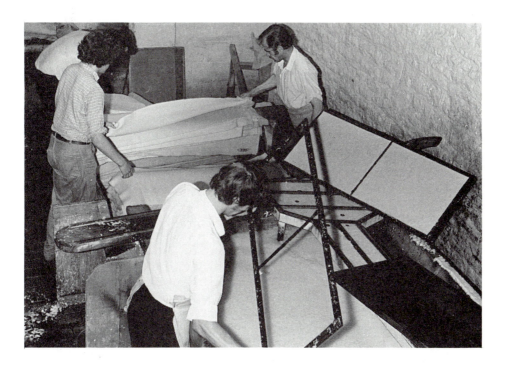

10. Vat team at Wookey Hole mill in 1970 using two sheet moulds. The vatman is about to place the deckle on the second mould ready for dipping while the other is draining on the ass. The coucher and his assistant are picking up a felt.

important point about the Hollander was that it could break down tougher rags and so dispense with the need for retting.

A book published in Holland in 1682 mentions beating rags with a roll and illustrations in the Dutch windmill books published in the early years of the next century show that by that time the Hollander had evolved into two types, one for breaking the rags with a second for beating them. It is tempting to see in some of the seventeenth century English patents attempts to introduce the Hollander here. For example Nathaniell Bladen's patent of 1682 claims,

> An engine, method, and mill, whereby hemp, flax, lynnen, cotton, cordage, silke, woollen, and all sorts of materials whereof all manner of pasteboard, and of paper for writing, printing, and for all sorts of other uses, hath been or may be made, are prepared and wrought into paper and pasteboard much speedier and cheaper than by the mills now used.[47]

Christopher Jackson made similar claims in another patent a couple of years later.[48]

It seems doubtful if any Hollander beaters were introduced until the early years of the next century. In 1725, Wansford Mill in Northamptonshire was equipped with four mortars and an engine and Bramshott Mill in Hampshire had both hammers and engines.[49] There were problems with their introduction because apparently people who were not properly trained tended to drive the roll too fast so the pulp was flung away from the roll. Then on early Hollanders, the bars on the bedplates were set parallel to the axle of the roll, but a better action resulted from their being set at a slight angle. In England, the Hollander was in general use by 1750 and by 1797 it was said that only one mill still employed stampers.[50]

The vatman's task in the actual forming of the sheet was made quicker, and possibly more pleasant by the addition of a heater or 'pistolet'. This is not found in southern Italy or Spain but is in regions further north. A copper drum was let into one side of the vat and a small fire kept burning in it to heat the stuff in the vat. The reduced viscosity given to the hot water made the pulp on the mould drain more quickly. In Scotland in 1793, William Scott, plumber, and George Gregory, tin-plate worker, patented heating the stuff in the vat by a steam coil which could be controlled by a valve to regulate the temperature.[51] At first, this was supplied with steam from its own boiler, but later, in mills with steam engines, steam could be taken from the main boilers. This device has continued in use up to the present day and should keep the stuff at a temperature of 80° to 95° F. One further addition to the vat was the mechanically driven agitator or 'hog' to keep the stuff continually stirred, relieving the vatman of another task. Again who invented this is not known but Joshua Gilpin saw one in William Simpson's Mill near Edinburgh during his tour in 1795.[52]

Another Dutch invention was the two-sheet mould which appears to have been invented in Holland around 1690. On it, a pair of sheets of paper could be made side by side and the deckle had a dividing bar down the middle. Normally these moulds were wider from left to right than from front to back because the laid lines and the watermarks on the sheets of paper had to run from left to right. The maximum stretch of the arms of the vatman was about 44 ins. (112 cm.) which limited the sizes of the sheets. These moulds became very popular in Holland and were introduced into North Germany in 1744. It is possible that Double Demy printing paper may have been made on such moulds in England from 1712 onwards. James Whatman the younger tried them at Turkey Mill in 1768, three years after he had visited Holland, and he may have been the first to use them in this country for making writing paper. Because making paper on them needed much more skill, the vatmen did not always greet them with enthusiasm but in the nineteenth century they became one of the ways of helping hand-made mills survive competition from machines.[53]

On the moulds themselves, ways of weaving the laid covers on looms instead of by hand appear to have been developed by 1700. In July 1694, Nicholas Dupin's petition to the English Privy Council mentioned that he could make moulds much faster than anyone else. An advertisement offering such a loom for sale appeared in a Germantown (USA) paper in 1744. In these looms, the chain wires were passed round either side of a guide which could be rotated through 180 degrees. All the guides were linked together so that, after the chain wires had been twisted together at the start, a laid wire was inserted across the loom between the chain wires by hand and then the guides turned one way to twist the chain wires and to hold that laid line in place. The next laid wire was inserted and the guides turned back again to secure that laid line in its position. Making a laid cover was speeded up considerably with such looms, and the covers were more even.

In 1795, Joseph Bramah patented an hydraulic press in which he claimed

> Certain New Methods of Producing and Applying a More Considerable
> Degree of Power in all kinds of Mechanical Apparatus and other Machinery
> requiring Motion and Force, than by any Means at Present practised
> for that purpose.[54]

He recommended its use in papermills for pressing out the water from the post of wet sheets of paper and felts because the pressure could be considerably greater than that of the earlier screw presses. The first in a papermill was almost certainly that in a mill started by Matthias Koops at Mill Bank, Westminster, where one press was installed by 1803.[55] In his patent of 1805, Bramah stated that one of his hydraulic presses could remove

> the necessity of employing such a considerable number of presses for
> the dry work, which is unavoidable in works of even but a tolerable extent,
> owing to the length of time the paper is obliged to continue in them in a
> compressed state, and on which account a larger capital is necessary as well
> as capacious buildings.[56]

59

11. The vertical shutters of the drying loft at Afon Wen mill, North Wales, in 1978 before they were removed.

He proposed one hydraulic press into which the post of paper could be wheeled on a platform on rails. These presses became common in most handmade mills later.

In order to dry the paper more quickly, heat was introduced into the drying lofts. In Scotland, the first recorded instance was in 1787 at Sauchie in Stirlingshire where there is a mention in an insurance policy of a 'Drying house Stove'. After 1795, the policies invariably mention such a stove or specify its absence. During

12. This computer controlled supercalender stack recently installed at Bowater's Sitting-bourne Mill is a far cry from the earliest type which consisted merely of a pair of wooden rollers.

his tour of Scotland in 1795, Joshua Gilpin saw at Simpson's mill at Polton a flue carried round the drying room to provide heat.[57] William Balston, when building Springfield Mill at Maidstone in 1807, installed steam heating apparatus for drying his handmade paper, one of the earlier uses of this medium which features on virtually all machines today.[58]

Finishing and smoothing the paper was speeded up when an Austrian smith, Hans Frey of Altenburg near Iglau, adapted the iron-maker's trip hammer in 1541 for glazing. This was operated by pressing down the tail of the hammer shaft, leaving the head clear with plenty of space round it. In 1761, de Lalande described one with an iron head weighing about 50 pounds with a face ten inches square. A few sheets of paper were put into a pack and moved backwards and forwards under the blows of the hammer. The glazing tended to be uneven and the blows weakened the paper.

Glazing hammers were replaced around 1720 by another Dutch invention, the glazing rolls or calender, which resembled a mangle. Later, metal rolls replaced the earlier wooden ones. The sheets of paper could be passed through either on their own or between plates of copper or zinc. The calender could also be used with heat. Either the rolls themselves were heated with hot irons placed in them or, more typically, the metal plates were heated and placed between pasteboards with the paper put between the boards. This later method is thought to be an English introduction around 1800. Baskerville, the famous Birmingham printer, used hot copper plates and passed them with the paper through rollers after he had printed on it which gave it an exceptional shine.[59] In 1790, Thomas Nightingale's patented friction calender introduced a new method for glazing. It combined iron and wooden rollers. The iron roller moved at a greater velocity than the wooden one and produced friction on the paper as it passed through, giving it a high gloss.[60] This principle was later applied to plate glazing.

Taxation

Paper was seen as a source of revenue by the Government from almost the beginning of its use and manufacture in this country. This account can only give a very brief outline of the major changes in a very complicated system of taxation on both imported paper and also that made at home. At least by 1507, all imported paper paid a customs duty of the usual 5 per cent *ad valorem* subsidy of tonnage and poundage and three types were mentioned. Between 1604 to 1660, eleven different types were enumerated which had risen to 67 in 1674 but this formed a totally inadequate list of the various types and sizes of paper in use. In 1712, new duties at a rate of 20 per cent *ad valorem* were levied which were charged according to a schedule covering some 37 categories of paper. The rates were raised in 1714 by 50 per cent in some cases and again in 1747, 1759, 1779 and 1782 when each time there was a 5 per cent rise. Pitt's tariff reforms hacked a way through this tangle of import duties and replaced them with a new set covering 56 types and sizes. Then in 1794 these duties based on value were repealed and the principle of assessment by weight introduced.[61]

The duties imposed on imported paper were heavier than the tax on paper of British make, so the British manufacturers had a measure of protection which gave them a decided advantage over their continental competitors in the home markets.[62] While duty was imposed on British paper during 1696–8, the first Act to impose an Excise tax on home-produced paper was passed in 1711 (10 Anne c. 19). In the ensuing 150 years, twenty-six other acts were passed which modified, extended and changed the incidence, the amount and the method of excising paper. This

Comparative Weight of Customs and Excise Duties, 1714 and 1752

| | 1714 | | 1752 | |
| | EXCISE | CUSTOMS | EXCISE | CUSTOMS |
TYPE	DUTY %	DUTY %	DUTY %	DUTY %
Pot, second	17	67	17	74
Demy, second	17	55	17	63
Brown	25	53	25	58
Brown, cap	10	39	10	44
Unenumerated (*ad valorem*)	18	60	18	65[64]

period may conveniently be divided into three sections, 1712–1780, 1781–1793 and 1794–1861. In the first of these, the tax was imposed at a certain amount per ream for eleven specified types of paper. For paper which did not conform to these specifications, a percentage of the declared market value was payable. This bore more heavily on paper of higher value. On any paper not included in the list, there was an *ad valorem* duty of 12 per cent which was raised to 18 per cent in 1714.[63]

In the second period, an attempt was made to assess the tax by reference to the physical description of all the papers on the market. The number of specified types was increased to 73 and non-conforming papers were charged at the next more expensive rate. The general weight of the duty was very much increased and it was soon made heavier still by further percentage increases in 1782, 1784 and 1787. The sheer complexity of the various tables for calculating the duty caused great difficulty for the manufacturers. Whatman's purchase of rags in the years 1780 and 1787 show that there was no great variation in his annual output of paper. Nevertheless, his Excise duties rose from £398 in 1780 to £2,831 in 1787, chiefly owing to new legislation.[65]

For the third period beginning in 1794, the tax on all paper was divided into five classes and assessed by weight. These duties were doubled in 1801. Though the duties themselves were simplified, the provisions for wrapping, inspecting, stamping and so on attained a new and fearsome rigorousness. The amount and class of paper were to be marked on the wrapper and numbered according to the number of such reams or bundles made at the mill in the current quarter. Each bundle had to be inspected, weighed and stamped by the Excise officer.

Customs and Excise Duties, 1794–1801

CLASS	DESCRIPTION	EXCISE DUTY per lb.	IMPORT DUTY per lb.
1	Writing, Drawing, printing, elephant, and cartridge	2½ d.	10 d.
2	Coloured and whited brown	1 d.	4 d.
3	Brown wrapping	½ d.	2 d.
4	Unenumerated	2½ d.	10 d.
5	Pasteboard, millboard and scaleboard	10 s. 6 d. per cwt.	20 s. per cwt.
	Glazed paper	6 s. per cwt.	12 s. per cwt.[66]

The general effect was to increase the Excise duties on all papers, as was to be expected in the second year of the Napoleonic Wars, but some encouragement was given to the manufacture of the best sorts of paper by increasing their duties only about 7 per cent, whereas inferior papers paid as much as 20 per cent more.[67]

In 1713, the gross amount of the Excise duty was £5,313 which increased to £8,722 in 1718. In the next few years, the amount raised rose and fell erratically until 1732 when the tax brought in £6,458. After this, there was a steady increase until £15,223 was reached in 1772. The effect of the Excise reforms of 1781 and the further rises in duty over the following years together with the reorganization of 1794 was to make the tax still heavier. Between 1770 and 1800, the output of paper rather less than doubled whilst the yield of the duty multiplied just over eleven-fold to £166,301 in 1800.[68] This was one reason for the tailing off in the expansion of the paper industry during the latter years of the eighteenth century.

V

THE WHATMANS AND WOVE PAPER

[They] may truly be said to have given additional smoothness to verse and a
new face to the literature of this country.

S. Ireland, 1793

During the eighteenth century, two British papermakers were to change the course
of papermaking history. They were the father and son, James Whatman. Not only
did they raise the standard of production in England until it rivalled and even
surpassed its continental competitors but they introduced new techniques which
created new types of paper. Their mill was Turkey Mill situated on the River Len
to the east of Maidstone.

George Gill had been making paper at Turkey Mill from at least 1680 and possibly 1676 or 1671.[1] In 1716 he was succeeded by his son William who went bankrupt
in 1729. When this mill was described by John Harris in his *History of Kent* in 1719,
it had become important because good quality white paper was being made there in
contrast with other mills in the region where only inferior paper was produced.

> There are some Paper-Mills also near it [Maidstone], which make a great
> deal of ordinary Wrapping Paper for Tobacco, Grocery Ware, Gloves, and
> Milleners Goods, etc. About Three-Quarters of a Mile *Eastward* of the Town,
> there is one which makes very good white Paper. 'Tis situated on what they
> call the *Little River*, which coming from *Leeds, Hollingbourne,* etc. runs into
> the *Medway* at this Place. Here it turns three over-shot Wheels, of about
> 8 Feet in Diameter, which moves the whole Work; the Water-Boards are
> about two Feet and a half long and the Trough delivers a Stream of Water
> of six inches deep. 'Tis a very large Work, and they could easily make much
> greater Quantities of Paper, if there were Demands accordingly. The Rags
> they use they have mostly from straggling Persons, which bring them to the
> Mill: And some they have from *London*. The Brown and White Brown Paper
> which they make here is chiefly from old Ropes, Sails, etc. and this Matter
> does not require about 12 Hours beating; but the fine Rags require 36 Hours

beating before they are fit to make White Paper. And Mr. *Gill* told me when
I went to see his Mill that he could not make his Paper fine and white, till
he brought into his Work a Collection of fine clear Water from two or three
springs which rise in a Field adjoining to the Mill.[2]

At this period, we would expect to find that this mill would be equipped with
stampers, which indeed is suggested by the time taken to beat the rags. Three
waterwheels indicate that there might have been three sets of these and so there
might have been three vats. If so, this was indeed a large mill at a time when most
others would have had only one.

In 1738 the freehold of Turkey Mill was conveyed to Richard Harris under
mortgage with two trustees, James Whatman and Thomas Harris, both described
as tanners. While James Whatman the elder had owned Hollingbourne Mill near
Maidstone since 1733 at the latest, there is no evidence that he was involved in
papermaking at that time because he let it to Richard Harris. Richard Harris was
carrying out an extensive rebuilding of Turkey Mill when he died in 1739. The
mill was left to his wife Ann whom Whatman married in the following year and so
became directly involved in papermaking.[3]

He started at a favourable moment for in 1739 began the war of the Spanish
Succession. Until that time, little good quality white paper had been produced in
England, but the war impeded trade with the Continent and supplies of high quality
printing paper in particular became scarce. By the end of the war in 1748, the
English manufacturers, including Whatman, had so much improved and increased
their output that they permanently secured a share in the market. In fact, Turkey
Mill became one of the largest, if not the largest producer of white paper by the
time of the death of the elder Whatman on 29 June 1759.

It is possible that Richard Harris was equipping Turkey Mill with Hollander
beaters when he died for they were being introduced into England at about that
time. In Italy, the surviving mills at Pescia and Amalfi show that the rags first were
cut up by stampers before being transferred to a Hollander beater for converting
into the whole stuff. In fact, nothing is known about the machinery that either
Whatman installed except that it has been estimated that later there were five vats.[4]
In the top of the mill was situated the drying loft. This was built across the valley in
the traditional British style with vertical sliding shutters so it caught the maximum
amount of wind. In 1787, John Rennie, the millwright, was engaged in installing
new shafting and wheels made of iron. He is described as 'substituting iron engines
for the traditional wooden ones' which could be referring to Hollander beaters
when ones made of wood were replaced with cast iron ones.[5] In order to produce
paper of high quality, the Whatmans must have kept abreast of the latest technical
developments in papermaking machinery.

The Whatmans were important first for raising the quality of white paper,
then introducing wove paper and finally for developing the use of wove paper. By

improving his equipment and also by carefully selecting his raw materials, the elder Whatman was able to manufacture paper particularly for writing that competed with continental imports. In China where calligraphy was practised with a fine brush, the paper could be quite soft. In the west, the use of quill pens necessitated a paper with a hard, strongly sized surface to prevent the tip digging in. Such paper Whatman produced, for from 1747 onwards, the State Papers (Domestic) files of the Public Record Office contain many documents written on paper that can be identified as his, as many at least as those of any other British manufacturer. His papers were much used in public offices and households of the great. In 1759, the year of the elder Whatman's death, they are found in letters written by the Dukes of Bedford and Dorset, the Earl of Pembroke and other peers. Almost all of them are in excellent condition and are considered to be the best on the market at this period. In fact this Whatman is credited with placing the English white paper industry on its feet. In 1747 the *London Tradesman* commented,

> We are but lately come into the Method of making a tolerable Paper, we were formerly supplied with that Commodity from France, Holland, and Genoa, and are still obliged to these Countries for our best papers . . . The French excel us in Writing-Paper, and the Genoese in Printer-Paper, from whom we take annually a great many thousand Pounds worth of the Commodity: However, our Consumption of this foreign Manufacture is lessening every Year, both on account of Interruption of Trade with the state of Genoa, and that we are now able to supply ourselves with large Quantities of our own Manufacture, little inferior to theirs, either in Colour or Substance.[6]

By 1748 and the end of the war, the English manufacturers, with Whatman leading them, had so much increased and improved their output that they permanently secured the English market.

One small improvement to writing paper probably was introduced by the younger Whatman when he took advantage of the even look-through of wove paper. Because note paper with printed lines was not available and in order to help people write in straight parallel lines, extra wires were sewn at regular intervals on top of the laid wires which shewed up in the paper and helped to guide the writer. This 'watermarked ruling' can be traced back to 1645 on the Continent.[7] However, by 1774, Whatman had applied the idea to his wove moulds with wires spaced at intervals of three-eighths inches.[8] In 1790, John Phipps, a papermaker of Dover, patented a similar idea for lined writing paper. Just in from one edge of a wove mould, he secured one wire to act as the margin and then at regular intervals down the cover he sewed wires at rightangles to the margin across the rest of the surface so they acted as guidelines for handwriting. Such paper continued to be produced for sixty to seventy years. Phipps also extended this idea to making outlines of sketches or drawings and saw this as a 'method to facilitate the acquirement of several of the useful and polite arts by giving an easy, effectual, and expeditious manner of teaching writing and drawing'.[9]

Turkey Mill was described in 1782 by Edward Hasted as having

> been much further improved by his son, *James Whatman Esq*; who with
> infinite pains and expense has now brought his manufactory of writing
> paper, for no other sort is made here, to a degree of perfection superior to
> most in this kingdom.[10]

This was not entirely true for, towards the end of the 1750s, some Whatman laid paper was being used by printers, but the only complete book, a quarto edition of Lucan's *Pharsalia* with Whatman paper of this sort, appeared from Horace Walpole's Strawberry Hill Press in 1760, the year after the elder Whatman's death. One reason for the fact that Whatman paper was not used by printers was that they needed paper with different characteristics from writing paper. Because the paper was damped before it was placed in the press, it was not sized so heavily in order to allow the water to penetrate it evenly.

Yet it was in the field of printing that the elder Whatman had made both paper and printing history. As the quality of printing improved, so people became dissatisfied with the irregularities in paper caused by lumps in the pulp as well as the different thicknesses caused by laid wires and shadow zones. De Lalande in his book on papermaking published in Paris in 1761 had commented on the adverse quality of paper with shadow zones but could not suggest how to remove them.[11] It was a problem which had faced the fastidious Birmingham printer, John Baskerville, who had devised a fount of type with very fine lines. To print with it, he required a paper as smooth as possible and it seems that he turned to James Whatman the elder. The final evidence for this comes from a meeting of Joshua Gilpin, the American papermaker, and James Whatman the younger in 1796 when Gilpin visited Turkey Mill and noted,

> His father, formerly a papermaker, . . . first made the wove paper in 1756,
> but he never brought it into repute till 1778 & 9.[12]

In March 1757, Baskerville published his quarto edition of *Publii Virgilii Maronis Bucolica, Georgica, et Aeneis*. While most of this book is printed on fine laid paper, some is printed on paper with no watermark and, what was then unique, with no laid or chain lines either. It has always been assumed that this paper must have been formed on a mould with a cover made from woven wire cloth, and hence the term 'wove' paper. Because Baskerville was a slow printer, the elder Whatman may have started to make this paper as early as 1754.[13] While the paper and sizing is of high quality, the paper is slightly marred by shadow zones showing where the ribs of the mould must have supported the covering. Presumably the wove cover

13. The wove cover of a Whatman mould

must have been placed directly on top of the framing of the mould, in the same way as contemporary covers for laid paper. The laid paper Baskerville used also shows shadow zones. Therefore it is believed that all this paper was made on moulds of traditional design.

However, in 1759, Baskerville printed on wove paper again for his edition of *Paradise Regained*, but, this time, the paper is even in look-through across the whole face of the sheet without any shadow zones or more translucent areas.[14] Baskerville chose wove paper for a third book in 1761, an edition of Aesop's *Fables*, as did Tonson in 1760 for Edward Capell's *Prolusions*. We know that Whatman made this paper because his son forwarded a sample of wove paper to the Society of Antiquaries in 1768 and stated that it was 'the same sort of paper I made for Mr. Caple's *Prolusions*'.[15] Yet this seems to have been the full extent of its use in this first period. Probably most of this paper was made before the death of the elder Whatman in 1759 and for some reason his son did not carry on with its production for a few years.

Four hand moulds may reveal how the perfect wove paper was achieved. One mould is in the Green collection at Hayle Mill, Maidstone, and is watermarked 'J. Larking 1817'. Larking was an early manufacturer of wove paper at East Malling

69

near Maidstone. The next mould in chronological order is dated 1826 and is preserved in the Manx Museum because it came from one of the mills on that island. Hertford Museum has some moulds from Hamper Mill where William Lepard was a rival to Whatman through making wove paper by at least 1792. One has a fleur de lys watermark and the date 1829. The final one came from Springfield Mill. It is a Whatman mould with the date 1831, now in the National Paper Museum Collection in Manchester by courtesy Whatman Reeve Angel Ltd. All these moulds have laid covers which are sewn directly onto the ribs and all have additional pairs of ribs or waterbars between the main ones under the chain lines.

It seemed possible that laid paper without shadow zones might have been made on these moulds. Therefore the staff at Hayle Mill tested this theory by making some sheets on the Whatman mould and at the same time others on a modern laid constructed in the traditional way without backing wires and without the additional waterbars. The paper made on the modern mould had shadow zones while that made on the Whatman mould did not. It is possible that the extra ribs were added originally to give greater support to the wove cover which would have been more flexible than a laid one. The fact that they also gave more even drainage may have been an accidental discovery, but this was probably the way in which making perfect wove paper was solved at first. It was an improvement which was soon applied to laid moulds as well.

During the eighteenth century, and particularly after 1760, the production of English prints produced from copper engravings increased dramatically as they met with approval both at home and abroad. Because no paper was manufactured in England which was suitable for their reproduction, in 1756 the Society of Arts started to offer two rewards, the first for making paper equal in quality to the French paper of the same sort which was agreed to be the finest, and the second for paper which would approximate most closely to the French.[16] Some of the features for which the Society of Arts was looking were described in a letter by one of their judges, Mr. Hadrill, a copper-plate printer.

> I have made trial of the different specimens of Paper received from the
> Society, and find upon the nicest observation, that that which is marked with
> five holes, is by far the best; I wetted it against the French paper, and found
> it takes the water equally well, and will keep much longer before it mildews,
> and is much superior to the French in cleanness: in this there is no Iron
> Moulds, which are very common in the French, and a very great defect.[17]

In this form of printing, the inked plate was placed on the bed of the press, the paper placed on top and then run under a roller, rather in the manner of an old-fashioned mangle. The rollers could impart considerable pressure through their narrow point of contact with the paper. The structure of the press, with its side frames, had sufficient strength to take the thrust from the parallel bearings for the ends of the rollers, which could be screwed down to increase the pressure further.

Sometimes the pressure distorted the watermark because the paper was printed damp. In these presses, it was possible to print sizes of paper much larger than in those used for ordinary letterpress.

Special paper had to be made which had to be strong enough to withstand this great pressure. Yet the other desirable features such as easy absorption of the ink tended to weaken it. Therefore plate paper was generally thicker than that used for letterpress printing. Complaint was made of English paper that, because the pulp had been prepared in Hollander beaters, the fibres were shortened which weakened the paper. Also English paper was too heavily sized. French paper, with its longer fibres did not need so much size to hold it together, kept its elasticity better, drew the ink out of the fine engraved lines on the plates and so rendered the print more brilliant. Paper for copper-plate printing needed to have an even surface, to be free

14. The lighter ribs on the back of this sample mould are the main ones with pairs of waterbars between them. This construction gave more even drainage and so avoided shadow zones in the paper. On this mould, a complex arrangement of backing wires can be seen between the ribs which also helped to give even drainage to a complex watermark.

of flaws and lumps and to be lightly sized. Although wove paper would have given a smoother surface, this feature alone would not have made the paper suitable for printing from copper plates. Softness was obtained by the addition of the more pliable cotton fibre to the linen furnish, which became a more common feature during the opening years of the nineteenth century.[18]

The second prize of the Society of Arts was awarded to various manufacturers, though not every year, for example in 1757 to Clement Taylor and in 1786 to William Lepard. In 1780 Lepard had taken over Hamper Mill, Hertfordshire. In 1787, J. Cary's *New and Correct English Atlas, Being a New Set of County Maps*, was printed on a fine creamy laid paper made by Lepard. In this instance, copper engravings were being printed on laid paper and it would be interesting to know if it were this paper which received the award. By 1792, Lepard was certainly making wove paper.[19] It was not until 1787 that the gold medal was awarded to John Bates of Lower Marsh Mill at High Wycombe, after which the premium was discontinued. We can probably find some of his paper in the second volume of the Supplement to F. Grosse, *The Antiquities of England and Wales*, published in 1787. In a corner of one sheet are the initials J B and, on the corresponding sheet after trimming, E S. This time the paper is wove.

It does not seem that either of the Whatmans ever submitted any of their papers for the premium of the Society of Arts. In 1768, James Whatman the younger offered to manufacture some wove paper for the Society of Antiquaries but none appears to have been produced. In his letter, Whatman made the comment, 'I must likewise know if it is to be sized of a proper strength for Printing only'.[20] If Whatman's trade had been principally in high grade writing paper, any made for printing would have been a special order. As far as can be ascertained, no wove paper was made again until 1773.

Yet James Whatman the younger had realised at least by 1768 that wove paper was 'infinitely better for copper plates'.[21] Its smoother and more even surface meant that the ink in the fine lines etched or engraved in copper plates would be transferred more cleanly and clearly onto the paper. In 1770, the Society of Antiquaries had commissioned Basire to produce a picture of *The Field of the Cloth of Gold*, illustrating the occasion in 1532 when Henry VIII of England and Francis I of France met between Calais and Boulogne. Unfortunately this picture turned out to be 49¼ in. x 27 in. (125 x 68.5 cm.), larger than any existing single sheet of paper. When sheets had to be joined together to make one of the correct size, the thicker double layer where they overlapped caused distortions in the printing and was visible in the finished picture.

On 7 December 1772, the Council of the Society of Antiquaries was read a letter from James Whatman offering to make some special paper for them.

> The Double Elephant which I at present make is 3 ft. 4 in. by 2 ft. 2½ in.
> [102 x 67 cm.], and is as large as any paper I have ever seen manufactured

in Europe. Two sheets of that pasted together would be large enough, but I suppose the bad consequences of that method is what they wish to avoid if a single sheet could be made. My present conveniences will not permit of my making any larger than the Double Elephant without alterations of most of the utensils, and even then it cannot be made by hand, but I have no doubt but a Contrivance I have thought of will enable me to make it, although that will draw on a certain expense of at least Fifty Pounds for things which cannot be of use to me on any other occasion.[22]

The Society of Antiquaries made further enquiries and found that the largest sheet of paper made in Holland, 47 ins. x 27 ins. (also given as 48 ins. x 27½ ins., 122 x 70 cm.), was smaller than that required by Basire. So Whatman was given the order which he had finished by November 1773. This paper was wove, which was always the type of Antiquarian paper made by Whatmans.

In fact Whatman was proved wrong about demand for this Antiquarian paper which was 53 ins. x 31 ins. (135 x 79 cm.) because 'there has since been a great demand from abroad, as plates can thus be worked off of a larger size than before was practicable'.[23] Whatman was clearly mistaken in thinking that his Contrivance, large moulds, bigger felts and press would be of no further use. He and his successors continued to make Antiquarian paper until about 1936. Although Whatman was exporting some paper from 1764, Antiquarian paper was amongst the earliest that was exported in large quantities from England, and its quality must have been able to rival continental manufacture.

The manufacture of Antiquarian paper was exceedingly costly because, instead of a vat team of three or four persons, eleven were required. The process of making it in the nineteen thirties was recorded. The mould, each one with its own deckle, was hooked onto rods suspended from Whatman's Contrivance, also known as the Bellows. The Bellows consisted of a pair of pivoted wooden beams, so that the Bellows Man at the far end could lower the mould into and raise it out of the vat. At either side of the mould were the pair of vatmen who had to work in unison dipping the mould, levelling it and giving it the shake to close the sheet. An assistant, the Pig in the Pound, helped the vatmen lift the mould off the hooks supporting it from the Bellows and pass it to the Ass Boy who held it vertically to drain. While the vatmen were attaching the second mould and its deckle, the Ass Boy and the pair of Deckle Men removed the deckle from the mould with the new sheet on it and passed it to the two Couchers. They placed the mould vertically on the edge of a felt, couched off the paper onto the felt, lifted it off and returned it to the Deckle Men. A post of 65 wet sheets and 66 wet felts was so heavy that it required six men (the Couchers, the Deckle Men and the Layers) to haul it to the hydraulic press. After pressing, two Layers separated the felts and paper, making a Pack containing 130 sheets. Antiquarian paper needed more pressings and time in the drying loft owing to its greater thickness. In Whatman's day, when there was no steam heating in the lofts, more than a year might elapse before the sheets were ready for the market.[24]

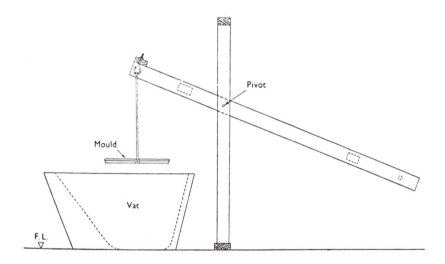

15. A diagram of Whatman's 'gallows' for making antiquarian paper, the largest size regularly made by hand.

The use of wove paper was to spread quite quickly for copper-plate engravings. For example, the Society of Antiquaries issued a series of prints of architectural subjects in the last quarter of the eighteenth century. The early ones were printed on laid paper but in Volume II of 1788, some were printed on wove and this became more frequent in Volume III published between 1792 and 1796. An even more important use of copper-plate engravings was in cartography. Detailed navigational charts were issued which required large sheets of paper. Damp conditions at sea would cause deterioration of glued joints where small pieces of paper had to be assembled to make large ones. Soon charts began to be printed on Whatman paper which ousted its Dutch rivals even on the Continent. At Naples, in the *Officina Topographica* just after 1790, the laid papers of Blauw and Honig made in Holland began to be replaced by Whatman wove which entirely superseded them by 1795. Then other English wove paper appeared in competition with the Whatman and British paper dominated that market.[25]

Although the elder Whatman and John Baskerville had pioneered the use of wove paper in books, it was to be many years before books became printed as a matter of course on this type of paper. In the last half of the eighteenth century, English papermakers faced great competition from the Dutch who were supplying most of the high quality paper used for printing. A survey of books printed about the history of Kent where good local paper was available has shown that even they were printed on Dutch paper. This is true of William Gostling's *A Walk in and about the City of Canterbury* published in Canterbury in 1777, as well as the first three volumes of the first edition of Edward Hasted's *History of Kent*, published respectively in 1778, 1782 and 1790, again at Canterbury. Even Francis Grosse's abridged edition of his *Antiquities of England and Wales*, published in London as late as 1798, was printed on imported paper.

On the other hand, a few sheets of wove paper do appear in Grosse's original editions of his work. Volumes I and II, issued in 1773 and 1774, are printed entirely on laid but a few sheets with copper plate engravings may have been used in Volume IV which appeared in 1775. The paper is very thick and heavy and the watermarking is difficult to distinguish. In the supplements issued in 1787, there is a great mixture of paper with some wove paper possibly made by John Bates (see page 72). William Camden's *Britannia* published in 1789 has a few of the engravings printed on wove paper in the first volume and in the subsequent ones this is used for most of the plates while the text is on laid.

James Whatman the younger played an important but not decisive role in helping to popularize wove paper in books. On 4 November 1786, Alderman Josiah Boydell gave a dinner in London at which George Nicol, the King's bookseller, suggested that Boydell should sponsor a folio edition of Shakespeare's *Dramatic Works*, to be published with the utmost typographical magnificence and embellished with engravings. It was a mammoth nine-volume publication filled with engravings by most of the well-known painters of the day, including Reynolds, West, Fuseli and Romney. This edition was 'particularly meant to combine the various beauties of Printing, Type-founding, Engraving and Paper-making'.[26] For this paper, Nicol turned to Whatman who of course recommended Whatman wove. When the first part appeared in January 1791, the excellence of the paper gave Whatman a great vogue in typographical circles, so much so that rival publishers began to plan similar huge folios, illustrated by contemporary artists.[27]

This fashion for Whatman wove paper soon spread among other publishers of smaller books of high quality. The Bible was printed by Thomas Bensley beginning in 1790, thus forestalling Boydell. Robert Bowyer issued Hume's *History of England* in sixty parts beginning in 1792. Others followed this lead. T. & J. Egerton of Whitehall published Samuel Ireland's *Picturesque Views on the River Thames* in 1792, followed by *Picturesque Views on the River Medway* in the year after. The latter is beautifully produced on Whatman wove paper throughout with exquisite illustrations. In fact, Ireland himself commented about Whatman,

> To this gentleman the country is much indebted for his great improvements
> in the art of paper-making, which he has unquestionably carried to a higher
> degree of perfection and excellence than was before known in this or
> any other kingdom, and may truly be said to have given additional

smoothness to verse, and a new face to the literature of this country.[28]

But printing on such thick and heavily sized paper could not have been easy when
each sheet had to be damped before being put in the press. One can well imagine
that the ordinary jobbing printer would not have liked Whatman's high quality paper
because it would have been more difficult to use.

While Whatman was leading the way in what might be called the 'coffee table'
market for illustrated books, other papermakers were producing wove paper for
printers. The earliest book discovered so far in which wove paper has been used
throughout is Richard Shepherd's *The Ground and Credibility of the Christian Religion*
published in 1788.[29] From that time on, it is possible to find a steady trickle of books
printed on wove paper. One of special interest is Job Orton's *A Short and Plain
Exposition of the Old Testament* because its six volumes were printed in Shrewsbury
in 1791 on an azure wove. Yet, for the greater part of the 1790s, wove paper was
confined mostly to the plates in ordinary books, possibly partly through its greater
cost. Examples of this are the *Evangelical Magazine* for 1793 which has the title
page and seven plates printed on unwatermarked wove, or Sprange's *The Tunbridge
Wells Guide* in which some of the illustrations were printed on wove paper made
by R. Williams with a date 1794. The same pattern is followed in John Banks's
A Treatise on Mills, published in 1795, which has three plates printed on wove and
everything else on laid paper.

However from the middle of the seventeen nineties, we begin to find the text of
ordinary books being printed on wove paper much more frequently. By 1797, its use
had spread to the north for a Sheffield printer published John Curr's *The Coal Viewer
and Engine Builder's Practical Companion* with the text on Whatman wove and some of
the plates on wove made by Whatman's rival, Lepard. The increasing popularity of
wove is shown by the second, small edition of Hasted's *History of Kent*, issued from
1797 to 1801 in twelve volumes, which were printed throughout on wove paper, but
not of Whatman's manufacture.

Matthias Koops would seem to sum up this trend. The first edition of his
Historical Account of Paper[30] was printed in 1800 on paper made from straw which
was produced on laid moulds throughout. There are four different watermarks, the
first the royal coat of arms with GR as countermark, the second the cypher GR
with 1797 beneath, the third 'Michaellet' and finally the letters NECKE. We know
from Koops's words that in May 1800 he established his own manufactory at the
Neckinger Mill in Bermondsey, just south of London. These watermarks may reflect
trials at mills other than his own and show laid moulds in common use. However, the
second edition of this work was issued in 1802 and was printed throughout on paper
made from waste paper. It is watermarked 'Neckinger Mill 1801' and is all wove.

Although he remained the most important manufacturer of wove paper in
general, James Whatman the younger soon faced competition. He himself decided

to retire in 1793 and, in the following year, had sold his mill to a partnership of William Balston, his former manager, and the Hollingworths, Thomas Robert and Finch. Balston continued with the Whatman tradition and the Whatman watermarks but, in 1804, decided to terminate the partnership and start a mill of his own. In the following year, it was agreed that the Hollingworths could use the watermark 'J. Whatman Turkey Mill' while Balston kept the plain 'J. Whatman' and the date. He later also used his own name.[31] Balston decided to launch out into a new era by driving his mill with a Boulton and Watt steam engine. The site he chose was just to the north of Maidstone by the side of the navigable River Medway so he had good transport facilities for bringing in coal and rags and despatching his paper to London. As the name of the mill, 'Springfield', indicates, his site also had its own supply of suitable clear water. The pulp was prepared by Hollander beaters driven by the steam engine but paper was made there by hand at first in 10 vats which was increased to 12 vats in 1860 and again to 16 in 1880. The tradition of handmade paper continued until a Fourdrinier machine was installed about 1930 and began to replace some of the hand production.[32]

In the important papermaking centre around Maidstone in the 1790s, there was soon established a group of manufacturers producing wove paper. Edmeads & Pine, at Ivy Mill at Loose just south of Maidstone, were making wove paper in 1791.[33] It is known that J. Larking at East Malling had followed by 1793.[34] The Taylor family at Shoreham a little further away are said to have made wove paper in the early 1790s and R. Williams of Eyehorne Street Mill, Hollingbourne had started by 1794.[35] Wove paper made by W. Elgar who controlled Chafford Mill has been found dated to 1796.[36] At the end of the century, there were J. Ruse at Upper Tovil Mill,[37] and Hayes & Wise at Padsole Mill, also at Maidstone.[38] J. Green, who started at Hayle Mill in 1810, was certainly making wove paper by 1815, if not from the very beginning. Watermarks which cannot be positively identified suggest that there were other makers in the area who entered this new market.

There was another important group of manufacturers to the north west of London centred around the River Wye in the High Wycombe area. During the eighteenth century, many of these mills started to produce good quality white paper. The success of John Bates at Lower Marsh Mill in winning the gold medal for copper-plate paper has been mentioned already. William Lepard at Hamper Mill must have been a formidable competitor to James Whatman, judging by the number of sheets of paper found with his watermark. He was certainly making wove paper in 1792[39] and for many years afterwards developed quite an extensive business. There may have been other early manufacturers in this region too but so far no watermarks before 1800 can be attributed definitely to any other mill here. After 1800, we find the watermark 'Rye 1807' which must have come from Rye Mill, High Wycombe.[40] Then there is 'G. Jones 1808', Griffiths Jones of Nash Mill, Hemel Hempstead,[41] which is close to this area and later became part of John Dickinson's. Finally

'Fellows 1815' is James Fellows, Soho Mills, again at High Wycombe which is now Thomas & Green Ltd.[42]

As far as England was concerned, the change from laid to wove paper for printing books was virtually accomplished around 1800 and this has remained true for the majority of books published ever since. That the change occurred about then is confirmed by two articles on papermaking. The first, in the *Sister Arts*[43] published in 1809, mentions 'moulds of brass wire, exceedingly fine and woven', as if they were a new invention. The second in Rees's *Cyclopaedia*, written around 1812, talks about a

> ... modern improvement with wire woven in a loom like cloth. This wove paper, as it is called, when made on these moulds, is a very superior article to the old paper, particularly for books, but a prejudice still prevails in favour of the old paper with lines, which obliges manufacturers still to make it, though by no means so fine or good as the wove.[44]

England was well ahead of other countries in producing wove paper. The next was France. Benjamin Franklin exhibited some wove paper in France in 1777 which made a favourable impression on papermakers and printers there.[45] The elder Didot, the noted Parisian printer and publisher, regretted that it was not possible to print on laid paper with some of the fine type produced by the English type-founder Caslon, and sent a letter to Johannot, a skilled papermaker of Annonay, asking him to copy Franklin's paper. After studying the paper of Baskerville's *Virgil*, he carried out some trial experiments with woven wire cloth and produced paper called 'papier velin'. The first French book to be printed actually on a pink wove paper was *Les Loisirs des Bords Loing* in 1784. The dates for the first production of wove paper elsewhere are Germany and the United States of America in 1795, Italy 1796 and Holland not until 1807. On 9th October 1807, the Dutch papermaker Jan Kool appeared before the notary Dirk Yff and told some fifteen of his colleagues how to make wove paper. The papermakers present had to pay a fee to Jan Kool and keep the manufacturing process a secret.[46] How quickly it became accepted in these countries is unknown.

The reasons for the introduction of wove paper for cartography and copper engravings are clearly that it had a smoother surface and that it did not vary so much in thickness. Why it became popular for books is harder to understand. Possibly the editions of Nicol and others after 1790 set a new standard that had to be copied. Then there was an element of chauvinism too. The wars with France after the French Revolution cut off supplies of paper from the Continent, including Holland. While there had been a steady decline in import of white paper to Britain throughout the whole of the eighteenth century, it was particularly abrupt after 1794. British production expanded more rapidly and was able to match the increasing demand.[47] The virtual cessation of imports from the

Continent certainly gave English manufacturers a boost and they not only met the challenge but seem to have exploited wove paper as an English invention. Koops pointed out in 1800,

> By perseverence, convenience in the construction of these manufactures, superior engines, presses and machines, and improved moulds, the industrious manufacturers have assisted and enabled to give English Paper its actual pre-eminence.[48]

VI

THE ORIGINS OF
WALLPAPER

The growth of the paper industry during the eighteenth century was stimulated in one section through the increasing popularity of wallpaper. Indeed it was partly to meet this demand in France that during the 1790s Nicholas Louis Robert turned his attention to inventing a machine for making continuous lengths of paper which would be suitable for decorating with wallpaper designs. Prehistoric paintings in caves show that, from time immemorial, people have wished to improve the surroundings in which they live, although we might not include graffiti in this category. The history of decorating walls also stretches back into the dim mists of antiquity. Remains in Egypt and elsewhere show that walls were often painted or plastered and painted. Such decorations were popular in the Roman Empire, and their public buildings were frequently adorned with murals and frescoes. Medieval churches presented a colourful appearance, even if some of the paintings portrayed horrific scenes of the last judgement and damnation.

For the domestic house, not only did the walls need brightening but, particularly in the northern parts of Europe, they needed an insulating layer to help keep the rooms a little warmer. Hangings such as tapestries or cloth fulfilled this function for they were literally hung by their top edges upon the wall as the background to a chair or something similar. To begin with, they seldom covered a whole wall but during the fourteenth century this became more common and by the fifteenth and sixteenth centuries tapestries were stretched over the entire surface, fastened at the four sides. They had the added advantage that they could be taken down for cleaning or removed elsewhere if the owner had to move home. So, as the ages rolled on, beauty of form and colour, protection against cold and damp, and fitness to display more luxurious tastes and culture, were the chief objectives in the use of wall-coverings.

One of the oldest and best-known tapestries is that preserved at Bayeux, showing William the Conqueror defeating King Harold of England. This was one type of tapestry where the pictorial design was embroidered onto a piece of existing cloth. There were also tapestries where the design was woven to become the actual structure of the hanging cloth. Much later, an Act of Henry VIII, dated 1512, mentions 4,000 pieces of tapestry as being imported in one ship,[1] and tapestries remained popular among those who could afford them until well into the 1700s. But they were expensive to produce owing to the time taken to sew and weave them, so they could be purchased only by the richer members of society.

A cheaper form of pictorial wall hanging was 'painted cloth'. We find reference to the excellent quality being produced in London as early as 1410.[2] It appears that the artists worked with water-colours on closely-woven linen saturated with gum water, the linen being laid on coarse woollen cloths which soaked up the moisture and prevented the colours spreading. We find references in Shakespeare's plays to 'painted cloth' and this material, being cheaper than tapestry, was purchased by a class for whom tapestry was too dear to buy. Here mention might be made of the later wall hangings where coarse canvas cloth was covered with a white surface on which patterns or pictures could be painted with an oil-based paint, similar to oil-paintings. Also, less expensive hangings were made from lengths of cloth with patterns woven in them. These patterns would be repeated many times in the length and breadth.

Just when paper was first used as a surface decoration to hang on walls is not known. It certainly has a long history of such a use in China. In 1481 occurs what still appears to be the earliest mention of the use of wallpaper in Europe. A payment, on behalf of Louis XI, was made to Jean Bourdichon for having painted fifty great rolls of paper blue, for inscribing them with the legend *Misericordias Domini in aeternum cantabo* and for colouring three angels, three feet high, to hold this paper at the word *Misericordias*. This paper was hung subsequently in several places at the royal Château of Plessis-les-Tours. The fact that this paper was in rolls implies that it must have been formed from separate sheets pasted together, something for which there is no other evidence until the latter part of the seventeenth century.[3]

The earliest paper used for decorating a room in England was found on the ceilings of the entrance hall and dining room of Christ's College, Cambridge, with a black and white design, 18 in. by 11 in. (458 x 280 mm.). Paper like this imitated woven cloth. When it was removed, it was discovered to have been painted on paper on the back of which had been printed a poem about the death of Henry VII in April 1509. Therefore this wallpaper must have been produced soon afterwards.[4] It was habitual to make wallpaper from paper which had been printed previously for books and so forth and had not been sold. This leads to a curious sidelight on the history of paperhangings and box linings of the seventeenth century. It would seem that this was the usual way of disposing of books that had offended against the political and religious opinions of those times. From the records of the Stationers' Company in 1673, we learn that the final end of censored copies of Hobbes's *Leviathan* was to serve this new purpose.

The main incentive for the introduction of wallpaper was its relative cheapness and the possibility of decorating it in such a way as to make a passable imitation of its more expensive rivals.[5] One method was by using a stencil and painting it. Plastered walls were often decorated in this way in the sixteenth century and, though the effect might not be quite so good, paper had the advantage of offering a dryer, less cold surface, as well as being portable. Early wallpaper was very much thicker than that made today. The transition from cloth to paper was gradual, but, already in the reign of Elizabeth I, there was a complaint that 'the painting on cloth is decayed.'[6]

Instead of paint, glue could be used and this led to the important popularizing of 'flock' papers during the seventeenth century. Coloured clippings of wool fibres from shearing cloth were scattered over the glued surface to give the effect of a richly patterned velvet or similar cloth. The patterns of glue might be stencilled over the surface to imitate brocade or damask patterns but François of Rouen, who is said to have been making flock papers soon after 1600, used wooden blocks, which had the pattern carved on them, for applying the glue to the paper. Some of these have borne the dates 1620 and 1630.[7]

Flock paper was the real forerunner of coloured effects in paper hangings as we know them today. The object aimed at, no doubt, was to satisfy the desire of gentle folk and well-to-do tradesfolk to adorn their homes with something comparable to the silk and velvet figured cloth hangings of the nobles. It would be a short step to use the wooden blocks intended for flock designs with a simple coloured pigment instead of the more elaborate flock effect, and so to produce wallpaper with coloured patterns. This would have given a sharper appearance than a stencil.

Some wallpaper, which cannot be dated, was found recently at Erddig House, Clwyd, showing an early method of manufacture. The sheets of paper are 23¼ in. wide and 17½ in. long (590 x 445 mm.). The pattern is printed beneath the over-lapping joins and so this paper must have been assembled after printing. The edge of the overlap which can be seen has been trimmed straight, but the underneath one still has its rough deckle edge. The paper is quite heavy and of reasonable white quality. It is 'virgin' and has not been used previously for something else. One of the most important innovations during the seventeenth century was the practice of pasting the sheets together and issuing them in rolls before painting. By this means, the horizontal joins of the paper might be disguised to some extent by the paint. It became easier to make large spreading designs, and less scope was left for the unskilled paper-hanger. The change seems to have started in the last decade of the seventeenth century.[8]

In 1680, George Minnikin, stationer of St. Martin's le Grand, was selling all sorts of coloured paperhangings and, ten years later, Edward Butling of Southwark was selling hangings for rooms in lengths and papers imitating 'flockwork, wainscott, marble, damask, Turkey work, etc.'[9] By the end of the seventeenth century, it may be said that, though the new craft by no means could be described as supplying a popular demand, there existed all the latent possibilities of the progress that was to come. In a general way, there had emerged three main branches of designs for wallpapers which have persisted ever since – one kind imitating figured textiles, such as brocades and damasks, a second imitating non-textile materials, such as marble, wood, leather, etc., and a third imitating pictorial decorations. These last will be discussed after seeing how paper was hung.

Even from the earliest times, some paper was 'hung', or stuck directly on the surface of the wall or ceiling. The country of its origin, China, probably started this fashion. Yet sometimes paper might be treated like a tapestry and hung from a beam or roller. In this case its removal was quite simple for it could be just rolled up. The paper was normally mounted on some sort of cloth backing. For hanging the best papers, a battened frame might be made. This would have been covered first with a loosely woven type of canvas, then with sheets of old newspaper, brown wrapping paper or cartridge paper and finally with the precious paper hanging itself,[10] which in this instance probably would have consisted of pictorial scenes. The canvas backing began to fall into disuse by the 1760s although the practice continued up to the First World War in some areas.[11]

Some of the earliest pictorial scenes came from China and had begun to be imported before the end of the seventeenth century. Ships officers of the East India Company were allowed to bring rolls of this paper home with them as one of their 'perks'. These scenes needed an entire wall to display them properly. In most of them, were to be found the characteristic Chinese 'landscape' motives, with foliage, birds and butterflies; others depicted customs and occupations of the people. A set invariably comprised twenty-five sheets, four feet wide and twelve feet long.[12] The pattern did not repeat so that the various lengths, when placed side by side in the correct order, presented a complete design spreading across from one to the other. Such papers were painted individually by hand for there was no way of printing them. Extra leaves or birds might be stuck on, as at Erddig in the State Bedroom, to help conceal the joins.

A regular trade with China sprang up and there is no doubt that the Chinese began to supply wallpapers which would appeal to the west regarding dimensions and general style but which retained their inimitable technique and artistic fluency. Soon genuine Chinese papers could not be procured fast enough, and English paper-stainers early in the eighteenth century began to produce papers in the Chinese mode. However these were easily distinguishable by the fact that, instead of the pattern covering the full wall space, it was usually repeated in squares or stripes and, along with a distinctively Chinese character of design, contained figures in European garb of the period, or, it might be, cases or other accessories such as the Chinese themselves never knew.[13]

Moreover, in the English imitations of Chinese wallpaper, the outline was usually printed from an engraved or etched plate, and the colour of the objects, often rivalling that of the real Chinese originals, was applied by hand. The genuine Chinese importations were hand-painted throughout, without, as far as can be ascertained, the help of any stencil or even an outline, for they had no repeats. The great period for Chinese papers was between 1740 and 1790, when all over England they are found replacing the decorations of earlier ages. One fine example dated to 1771 is in the State Bedroom at Erddig.[14] The paper on which these Chinese designs were painted seems to have been produced in much larger sheets than was customary in the west. At Erddig, the width of the strips is roughly 37 in. (940 mm.). The horizontal overlaps are more difficult to distinguish but they occur at about 21 in. (533 mm.). This paper would have been mounted on a backing or lining paper. In another instance, the sheets of this secondary paper, which was made from bamboo, measured 48 by 36⅞ in. (1220 x 937 mm.); a size much larger than was being produced in Europe.[15] In 1792, wallpaper imported from China was no longer exempt from tax.[16] In any case, popularity of the Chinese designs had begun to wane, partly because they lacked depth and made no attempt to disguise the two-dimensional nature of the wall.[17]

The first English patent specifically for paper hangings was granted in October 1692 to William Bayley. Wallpaper became more popular after Jean Papillon in France soon after 1700 developed patterns which repeated over a number of joined sheets, even to the extent of using wood-cut blocks three feet in length.[18] Then, around 1730, the architect Kent was commissioned by George I to redecorate Kensington Palace and he chose to hang paper in the Great Drawing Room. This was the first time at this period that wallpaper had been used in a royal building anywhere in Europe and the effect was greatly admired.[19] The use of wallpaper for wallhangings increased dramatically throughout the eighteenth century, through the generally increasing prosperity caused by overseas conquests and trade, the agrarian revolution and then the Industrial Revolution, all of which created a middle class who could afford wallpapers. In addition, throughout this period, there was a steady increase in population. In 1700, the population of England and Wales was over 5,000,000 which had doubled by 1801. The most rapid expansion came between 1801 and 1821 when there was an increase to 12,000,000.[20] The next doubling took less than sixty years and in 1911, there were 36,000,000 people in England. Between 1801 and 1911, the number of homes in England had risen from 1.6 to 7.6 million. This shows the potential for the supply of wallhangings, which could be met only by mechanisation.

Such a trade did not escape the attention of Government officials looking for sources of revenue. By a statute of Queen Anne in 1712, a duty of 1 d. a square yard (increased two years later to 1½ d. per square yard) in addition to the ordinary duty on paper was imposed on stained paper, e.g., wallpaper.[21] In 1773, the ban placed on the importation of foreign painted papers in the time of Richard III was expressly repealed but a customs duty was imposed instead, in order to balance the Excise duty on English papers. The papers of the East India Company, e.g. the Chinese designs, could still be imported free until 1792, when the basis of the customs duty on all imported wallpaper was changed to 6 d. per pound weight in addition to any duties on paper. In the meantime, the Excise duty on English paper hangings had been increased from 1½ d. to 1¾ d. in 1777. In these years, wallpaper tax receipts were:

	£
1770	13,242
80	11,955
90	19,204
1800	24,811
10	32,228
20	34,246
30	44,835
1833	53,986
34	63,795[22]

The 1714 Act taxing British wallpaper was repealed in 1836.[23] The customs duties on foreign imports were raised to 9 d. per pound in 1803 and an extra 3 d. was added in 1809. It was not until 1847 that they were reduced to 2 d. and then 1 d. six years later before finally disappearing in 1861. The duty on paper itself was abolished at the same time by Mr. Gladstone as 'a hinderance to education and a tax on knowledge' – a step which meant an immediate cost to the Exchequer of £1,200,000 a year.[24] The Excise figures show the scale of wallpaper manufacture in this country. The output of paper hangings in 1770 was 255,731 pieces and in 1834 1,222,753. The increase from then onwards was startling, due partly to the freedom from taxation and due partly to the development of rotary printing. In 1851, the production was 5,500,000, in 1860 19,000,000 and in 1874, 32,000,000 pieces.

Customs House Records 1774

	sq. yds.
Channel Islands	7,060
Ireland	2,471
Flanders	96,230
France	2,800
Germany	27,521
Holland	101,295
Italy	32,946
Portugal	110,612
Russia	21,453
Scandinavia & Eastern Baltic	9,184
Spain	171,859
Africa	350
British West Indies	23,994
Canada	26,963
Florida	2,842
Nova Scotia	294[26]

A great deal of this production was exported. Even by the 1750s, English products of flock papers were unrivalled both artistically and technically. In 1768, Papillon's son, Jean Michel, joined in the general eulogy of English wallpapers,[25] and there was a great interest in them in France, so much so that some English manufacturers established themselves in Paris. In 1774, the Customs House records show the extent of the English export trade.

From these figures, it may be seen that the industry was in a flourishing state, yet there is one area that is missing, the country which soon became the United States of America. After independence, the United States turned to France to supply her with wallpaper and this gave the French manufacturers their chance. In 1787, the consumption of French paper-hangings in the United States was so great that the French Government took off the export duty.[27] In the 1790s, the quality of the French designs began to surpass that of the English. France was quite well aware of the secret of her superiority. In Anno VI of the 'Revolutionary Calendar' (1797), the Jury of the Lycees des Arts reported (translation):–

> Our manufacturers of paperhangings have reached a high degree of
> perfection. France owes her superiority in this respect to the study of
> drawing, which ordinarily forms part of the education of the industrial classes,
> and of which the knowledge has spread to the well-to-do, who are the prime
> consumers, and whose taste ultimately determines the direction which the
> manufacturers give to their work.[28]

Soon wallpaper was no longer the poor relation of other decorative materials and had a role of its own. The number of wallpaper businesses in France grew from fifty in 1789 to sixty-seven in 1803 and ninety-six by 1811.

There was also a change in style, starting in the 1790s, for the period between 1800 and 1860 was the time when scenic or panoramic wallpapers became the vogue. Their designs were compared to fine art and had become immensely popular by the early part of the nineteenth century. Some of the smaller designs were printed in colour on panels that measured 8 to 10 feet (2.4 x 3 m.) in length and contained large non-repeating pictorial illustrations. There were also panoramic wallpapers, comprising up to thirty individual lengths whose first and last panels adjoined to form a continuous scene that covered the walls of an entire room. Both types contained scenes that were not only representational but highly illusionistic in effect and employed devices such as perspective and modelling which created an impression of space and depth utterly at variance with the flat and solid surface of the wall.

Their production was time-consuming and thus costly. The designing, printing and engraving required much careful planning and considerable experience and skill on the part of those involved. First, a subject was selected and an artist was asked to submit sketches indicating the composition and the divisions within the overall design. These were then enlarged to full-sized cartoons, which showed the distribution of the printing blocks. The difficulty here lay in attempting to translate a painting, with all its subtle gradations of tone and detailed lines, into a medium such as wallpaper which relied on flat areas of colour to achieve a similar effect. The designer needed to have a close understanding of the different processes involved in block printing and sometimes someone other than the original artist might have to design the cartoons.

Once this stage was completed, the preparation of the blocks began. They were similar to ordinary calico printing blocks. Each was made of three or four layers of wood and measured approximately 18 by 20 in. (457 x 508 mm.). The design was first traced and then engraved onto the surface of the block, which was made of some particularly fine-grained material such as pear or sycamore. For every colour and motif, it was necessary to engrave a separate block and in some decorations, where over two hundred colours were used, their number was extraordinarily high. In 1855, a French paper-hanger was engaged in producing a design which required over 3,000 blocks.[29] The printing was all done by hand and the printer worked to an exact order, with water-based paints, laying down the darker colours and broader areas first, and filling in the lighter tones and details later. Each layer had to dry before the next colour could be applied and decorations of twenty colours or more would take several months to print. The total period of time involved in preparing and completing a scenic wallpaper has been estimated at between eighteen months to two years, but this did not deter ambitious manufacturers and a surprisingly large number was produced.[30]

The accuracy of the printing in these scenic wallpapers, and others too, depended upon a device developed in England in the middle of the eighteenth century. In the earliest designs with more than one colour, there was not much trouble with 'register', e.g., seeing that the block of one colour was placed accurately over the position of the previous one so that the patterns mated properly, because the patterns were mostly geometrical and so the various parts could be fitted with reasonable accuracy. The methods of printing with hand blocks used up to the present day were established a little before 1759, when Robert Dossie published his *Handmaid to the Arts*. Each block was fitted on the sides with locating or 'pitch' pins which printed a dot at the edge of the pattern. The pins on the next dot were lined up accurately on top of the dots of the previous colour left on the paper. It would seem, however, that it was not until circa 1770 that enough experience had been gained in the use of this medium for workmen to do justice to the designs with which they were confronted. While the new process was still being perfected, the use of stencils was not very much affected. It was not until a comparatively late date in the century that competition from printing in colour from wood blocks became really serious and stencilling became restricted to the cheaper grades of wallpaper and flock papers with bold designs[31].

Until 1840, all wallpapers were produced by hand. In printing with hand blocks, the paper was laid out on a long table and often first given a wash of background colour. To 'ink' the pattern blocks, the colour was spread upon a felt in a large tray and the block placed face down on top of it. The block was lifted off and aligned on

16. Carved wood blocks for printing wallpaper fixed on the end of a barn near Marsh Mill, Henley, in 1978.

top of the paper by means of the pitch pins. On cloth printing, the back of the block was hit with a hammer to make the colour penetrate. On wallpaper, the blocks were pressed down with long levers operated by boys swinging on their ends.

There seems to be no information about who prepared the rolls of paper on which the designs were printed. The length of a 'piece' of wallpaper corresponds to the length of a piece of woven cloth. At about twelve yards, the length was far too long to be made in a single sheet by any hand papermaking techniques so it had to be assembled from smaller ones. In the very early period, the 'pieces' of wallpaper may have been produced from sheets of paper of varying sizes, as is shown in an example dated to 1680 in the Victoria and Albert Museum. Soon, however, each piece was assembled from sheets of paper all the same size, but there was no standardization in these sizes. By the middle of the eighteenth century, wallpaper was being sold in a standard width of 22½ in. (571 mm.) (although examples seen range from 22 to 23½ in. (558 to 596 mm.)) and in lengths of about eleven and a half yards. On examples made even in the early 1820s, the length of each individual sheet might be only 16 in. (406 mm.)[32] Generally, the paper was made on a laid mould, and the laid lines, except in one example, ran across the paper, e.g., they were horizontal when the paper was hung. Wallpaper produced in the 1820s was more likely to have been made from larger sheets, either 'Elephant' (22½ in. wide by 32 in. (571 x 813 mm.) long) or more usually 'Double Demy' (22½ in. wide by 35 in. (571 x 889 mm.) long), so thirteen of the former and twelve of the latter made up a piece.

The sheets were usually joined by overlapping from ⅜ in. to ½ in. (10 to 12 mm.). The rough deckle edge was generally trimmed off to leave a straight line. The joins could cause problems with printing for the thick ink might not penetrate to the bottom layer of paper, leaving an unprinted area. For better quality sheets, for example those used in copper-plate printing, the edges of the sheets might be

17. Printing wallpaper with hand blocks. The boy swung on the end of the lever to give additional pressure.

89

18. A twelve colour wallpaper printing machine, c. 1860.

thinned down and tapered so the joins were smoother. The joins were often visible when the paper was hung on the wall. It would be interesting to know whether they came unstuck, either through the dampness when the paper was being printed, or when the glue or size was applied when it was being hung. The vertical joins were butted together but first it was necessary to trim off both side edges, or selvedges, from the rolls before hanging, a task which had to be done with some accuracy.[33]

There were many reasons for having wallpaper made in continuous sheets and this was probably one of the incentives behind Nicholas Louis Robert's invention of the papermaking machine in Paris in 1799. But this machine was never brought into production until it had been considerably modified and improved in England by the Fourdrinier brothers, who had their first one working in 1805. However, the Excise authorities in this country would not permit continuous lengths of paper to be printed and insisted that they be cut up because the paper was still taxed by the sheet. Some of the larger sheets found in the wallpapers of the 1820s have been made on wove wire and, from the shorter nature of the fibres in their texture, may well have been made on a machine. Also some papers of this date are considerably thinner than earlier ones and these too may be machine-made. One well-known London firm, Messrs William Coopers & Co., who had a large wallpaper factory in West Smithfield, on three successive occasions before 1828 was refused permission by the authorities to use continuous paper. In 1825, the ban on the importation of foreign wallpaper was removed and French papers began to enter the country despite the almost prohibitive duty of one shilling per square yard. This forced the Excise authorities' hand who permitted the use of continuous paper in 1830.[34]

This paved the way for printing wallpaper by rotary printing presses. In 1764, Thomas Fryer had invented a printing machine with the design engraved on cylinders but it was not a success.[35] Although, in England, foot-operated levers had replaced their hand-held counterparts by the 1850s, conditions in the industry remained quite arduous for adults as well as young boys. The heat from the drying rooms, where the papers were hung, was often intolerable, while dust from the flocking process was likely to cause asphixiation! One historian cites a government inquiry of 1832 which found that, by the age of fifty, the majority of block printers were either too debilitated by respiratory complaints to continue working or had succumbed to disease brought about by over-exposure to the large quantities of lead and arsenic mixed in the wallpaper paints.[36] Green generally contained arsenic which was later banned.

The first successful machine was patented in 1839 by C.H. and E. Potter who owned a cotton printing firm based in Darwen, Lancashire. The paper passed round the surface of a large cylindrical drum and received an impression of the pattern from a number of rollers arranged around it. The printing rollers were inked with colours stored in troughs beneath each one. At first, engraved copper rollers were used but, in 1840, Potters turned to wooden rollers with raised patterns and put the first commercially acceptable wallpapers on the market in 1841.[37] Wallpaper printing machines have remained in this form up to almost the present day.

Progress was quick for by 1850, machines capable of printing perfectly registered designs in up to eight different colours were common.

> It is most satisfactory and surprising to witness the rapidity and precision with which papers of 6 or 8 colours are run off, the whole eight colours being printed during the passage of the papers *once through the machine.* A single machine is capable of printing in one hour 200 pieces of paper, each 12 yards long, or 1,500 pieces equal to 18,000 yards or 54,000 feet per day,[38]

At the Great Exhibition in 1851, specimens of machine-printed paper were shown with fourteen different colours and others with twenty colours made by fourteen rollers.[39] The effect of printing machines upon production was staggering, for, from 1,050,000 yards in 1834, it rose to 9,000,000 in 1860. It was estimated that in 1851, 5,500,000 pieces were produced valued at £400,000. Prices dropped to as little as ¼ d. a yard. Within just one generation, wallpaper, previously a luxury item, became a commodity available to all but the very poor.[40]

THE DEVELOPMENT
OF THE FOURDRINIER
PAPER MACHINE

The ancient traditions of making paper by hand were based on craft techniques which meant that production occurred in small batches. Centuries old methods still predominated in 1800 for all types of paper, whether it was high quality writing paper, less good paper for wall hangings or coarse wrapping paper. Certainly in England, no attempts had been made to develop any machines for forming the sheet but people had been trying on the Continent. Three different approaches were followed. One was to mechanise the handmade process by forming sheets of paper on individual moulds. Another was to use a drum covered with either laid or wove cloth to act as a strainer. The drum either might be immersed in a vat of stuff or the stuff might be poured onto its surface. In either case, the water was strained through the drum, leaving the sheet of paper on the surface which could be drawn off in a continuous sheet. The third way was to pour the stuff onto a belt of wire cloth moving continuously so that the water drained away through the cloth, once again leaving the sheet on the surface. It was this last method which was brought first to a commercial success.

Nicholas Louis Robert (1761–1828)

Nicholas Louis Robert was employed by Didot Saint Leger in his paper mill at Essonnes, twenty miles south of Paris, as clerk inspector of workmen. Robert quickly became aware of the difficulties in controlling three hundred people whose minds were inflamed by the progress of the French Revolution. Lack of discipline among the men disrupted the production of paper for the 'Assignats' or banknotes. Encouraged by Didot, Robert must have started to make a model of a machine for papermaking, possibly in 1793, but it did not succeed and was laid aside. He tried again and was certainly experimenting in 1796. Within two years, he had a model which performed so satisfactorily that Didot was encouraged to order the carpenters

19. A model of the Robert papermaking machine constructed from the drawings in 1984.

and smiths in his mill to help make a full size machine. In a few months, a machine was completed, capable of making paper to the width of Colombier (24 in., 61 cm.) and of various lengths. This was the width of wallpaper then used in France and the manufacture of lengths of paper for printing wallpaper seems to have been one of Robert's objectives.[1]

In September 1798, Robert asked that the normal fee for a patent should be waived because he could not afford it but this was refused and instead he was awarded a grant of 3,000 francs. This enabled him to apply for a patent costing 1,562 francs which was granted on 18 January, 1799.[2] Didot offered to buy the patent from Robert but the conditions could not be settled between them. Robert tried to start making paper on his machine by himself but he did not have enough capital. Eventually Didot claimed the machine as he had provided the materials for its construction. Finally, a reconciliation was effected but Didot never paid the balance of the money agreed.

In the meantime, Didot was anxious to take out a patent in England where he thought prospects for developing the machine were better. As he could not leave

France, he proposed that John Gamble should go to England with the drawings and samples of paper produced on the machine. Gamble was Didot's brother-in-law and was an Englishman working in Paris on the exchange of prisoners of war. Gamble had connections with the wallpaper industry for he had a brother who lived in Paris at the time of the Revolution and was described as a papermaker and engraver.

Gamble's superior officer, Captain Coates, gave him letters of introduction to the Mayor of Dover, one T. Mantell, Esq., where he met Mr. Millikin. When Millikin heard about the papermachine, he undertook to introduce Gamble to the Fourdriniers, wealthy stationers in London. After the Fourdriniers had seen the nine or ten rolls of paper made on Robert's machine, which Gamble had with him, they agreed to purchase a share in the project and to provide the finance. In 1801, John Gamble applied for an English patent for an 'Invention for Making Paper', which was enrolled as Number 2,487 on 20 October that year.

The Patents

The original texts and drawings of both Robert's and Gamble's patents have survived in Paris and London.[3] In addition, five of the six drawings which John Gamble took with him from France to England, are now in America, owned by Dr. L.B. Schlosser. The complete set of six drawings consists of:

1. Lower plan showing the framework at the bottom of the machine.
2. Upper plan, a view looking down on top. (This one is missing from Schlosser's collection.)
3. A side section through the middle from one end to the other showing the pulp or 'stuff' in the vat.
4. Side elevation from end to end.
5. Two drawings, one a cross-section taken by the press rolls and an end elevation from the 'dry' end.
6. Two more cross-sections, one through the middle of the vat and the other actually through the press.

All the drawings are coloured, including shadows, which helps to give a three-dimensional effect. The main frame is painted a light straw, with a brownish grey used for the shadows and also for the vat. Where there is a cross-section, the ends of the cut pieces of timber are painted pink. Copper or brass is coloured yellow. All the iron-work is dark blue, while the stuff in the vat is pale blue. The felt covering the rollers is white. It is interesting to note that these colours remained standard on engineering drawings for another hundred years.

These drawings are not dated. They have on them signatures of John Gamble, Bloxham and the Fourdriniers, but these could have been added when agreement was reached to form the English partnership. The surviving drawings in France may have been copied from this set while those in the English patent, although much smaller, definitely were. There are scales in both 'pieds' and 'metres' on all sets which vary from drawing to drawing. Then there are minor differences such as the letters which identify different parts. In the English patent, all the drawings were copied onto a single sheet of parchment or vellum, 40 in. long (roughly 102 cm.) by 17 in. high (roughly 43 cm.). Each illustration is about 11 by 6¾ in. (28 by 17 cm.) and is an exquisite miniature version of the large ones.

As was customary at the time for English patents, the text was written on sheets of parchment, sewn together and then rolled up. Gamble's patent starts a new roll so it is right in the middle and the title has become covered with dust over the years. The drawings are enrolled with the text too. There are in fact two sheets of drawings because the patent starts by describing a second dry press to be added to the original machine. There is a first sheet of drawings illustrating this press and also an agitator to be added to the vat of the paper machine.

After the description of this first sheet of drawings, comes the description of the drawings of the paper machine itself. The text is a direct translation of the French, but the way it has been laid out has caused errors when it came to be printed in 1856. The French patent describes the drawings sheet by sheet, with a heading 'Feuille 1ere', and sub-sections for each figure. The identification letters, e.g., AA, BB, are set clearly to one side, with the description alongside in paragraphs. Now English regulations decreed that it must be impossible to alter the text of a patent, so the English version was written out in a solid mass, without any paragraphs or breaks. It would be difficult enough to determine where the description of one sheet of drawings ends and the next begins, but the word 'Feuille' has been translated as 'No.', i.e., number. The result is that, in the printed version of 1856, the descriptions of sheets 2, 3 and 4 start in the middle of paragraphs!

It is evident that the English translator was not familiar with papermaking terms. For example, the French 'Cuve' has become a 'vessel' and not 'vat'. In sheet 2 ff., 'Liens de fer' has become iron chains and not bands or straps. More important, the wire covering on which the sheet of paper is formed is called 'toile de cuive' in the French, but 'sheet of copper' in the English. However, the word 'cloth' is used at sheet No. 2, NN, where the tensioning device is described. The present term in French to describe a 'laid' covering is 'verge' so it cannot be determined whether 'toile de cuive' means a 'laid' or a 'wove' wire.

Later Patents

Although sample sheets of paper were made on the first machine in France, production was never very satisfactory, and, with hindsight, we can see some of the reasons why. The critical examination of this machine which now follows also takes into consideration two further English patents which show how the original was improved and gradually changed into the Fourdrinier machine we know today. The next patent is No. 2708, enrolled 5 December 1803 in the name of John Gamble, for certain improvements and additions to the machine described in his 1801 patent. This is clear from references in the 1803 patent to the drawings and identification letters in his earlier one. The final patent is No. 3068, enrolled 13 February 1807 in the names of 'Henry Fourdrinier and Sealy Fourdrinier of Sherborne Lane, London, Paper Manufacturers, and John Gamble of St. Neotts, in the county of Huntingdon, Paper Manufacturer'. It shows a machine which can be easily identified as a direct ancestor of the present Fourdrinier paper machines, with most of the essential features there, laid out in a similar way. Missing parts are the vacuum suction on the wire to remove the water and also the drying cylinders. Additional reference will be made also to two drawings by Bryan Donkin which have been dated to 1803 and 1804.[4]

Robert's original machine was brought to England in 1802. It was just over 8 ft. long by 4 ft. 6 in. wide (2.4 x 1.36 m.). As well as being stationers, the Fourdriniers were also involved in papermaking for they had purchased Two Waters Mill, Hertfordshire, in 1791.[5] Their millwright or engineer was John Hall of Dartford and they turned to him for help with improving Robert's machine. Didot came over from France and with Gamble superintended the alterations being made. Hall felt he needed more assistance and sought help from Bryan Donkin, a former apprentice of his who by then had his own business as a mould-maker and who was also an extremely ingenious and clever mechanic.[6] In 1803, an improved model based on the French original was erected for trials at Frogmore Mill, also in Hertfordshire,

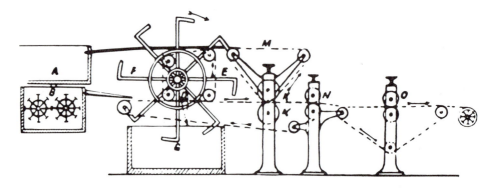

20. By 1803, Bryan Donkin had considerably improved Robert's machine by adding a breast box with agitators on the left, a top wire (M) on the first press and a second couch press (N). The paper was transferred to an endless felt and passed through a third press (O) before being reeled up.

which the Fourdriniers had purchased for that purpose but the paper produced on it proved to be unfit for sale. In the meantime, Hall, who does not seem to have been particularly interested in the papermachine, gave his consent for the Fourdriniers to establish a factory for Donkin at Fort Place in Bermondsey on the outskirts of London where work could start on another machine which in its turn was taken to Frogmore Mill in 1804. This machine was 27 ft. long by 4 ft. wide (8.2 x 1.2 m.). The evolution of these machines shows how Robert's original one had to be altered in order to make paper on it successfully.

The Papermaking Machine

The Vat

It is proposed to describe the machine by following the course of the pulp through it and seeing how it would have been turned into a piece of paper. Robert constructed his machine around a large oval vat. At one end, he placed the mechanism to raise the pulp and pour it onto the beginning of the endless wire on which the paper was formed. The water strained through the wire and ran directly into the vat below and the same happened with the water squeezed out by a pair of press rollers placed at the further end of the vat. The wire extended over the far end of the vat where the paper was taken off and wound onto a removable couch roller. Robert envisaged making sheets 'from one to twelve feet wide' so that his vat would have been an enormous size if he had tried to make paper of the latter width.

The vat was made in the same way as those shown in the many eighteenth century illustrations of hand papermaking mills and was a neat example of the cooper's art, with vertical staves banded together with three iron hoops. Neither the French patent nor that of 1801 say how it was filled. If the way papermaking by hand which survives at Ambert in France is typical of eighteenth century methods, then the vat was filled with fibres only at the beginning of making each post. The vatman compensated for the diminishing consistency of his pulp as he took out each sheet of paper. If the vat on the machine also was filled only at the beginning of each make, then the sheet must have become thinner and thinner as it was being formed because the fibre content in the vat would have fallen rapidly. The machine could not have adapted itself in the same way as the vatman to the thinner pulp. Therefore, either more stuff had to be added continually, which does not seem to have been done because none of the drawings shows any replenishing pipes, or the volume of the pulp being poured onto the moving wire should have been increased steadily. But the gear ratios are fixed and could not be varied and so thickness of the sheet of paper must have diminished as it was being produced.

It is possible that Bryan Donkin found the correct solution to this problem in 1803, when the vat was divorced from the rest of the machine. His first drawings show a vat or breast box at one end out of which the pulp flowed over a ledge and by gravity down a slope onto the wire. By the time of the 1807 patent, the rate of flow of the pulp could be controlled finely, but it was not until 1808 that we have definite evidence in another drawing of the final, essential component, the stuff chest.[7] Thicker pulp could be stored in the stuff chest and run steadily into the breast box where it could be diluted to the right consistency. Then a constant stream of pulp of regular consistency could flow onto the wire.

On the Robert machine, not only would the pulp in the vat have become thinner as the paper was being made, but the level would have fallen. As it is depicted, the lifting drum soon would have failed to scoop up the pulp. An attempt was made to rectify this in the 1803 patent by having two drums, a small one on top of a lower one, with the smaller one above the surface. These became redundant when the vat was moved to the end and the pulp run out by gravity.

In the early French mills, the vatman had to stir the pulp by hand. None of the early drawings show any method of keeping the stuff in suspension and Robert must have found that the fibres quickly dropped to the bottom of the vat. On the first sheet of drawings of the 1801 patent, figure 7 shows an agitator in the vat, driven by gearing probably off the pulp lifting drum. There is nothing to indicate whether this was invented in France or England, but it has remained an essential feature in paper machine vats and storage chests ever since and has been adopted in hand mills too.

The Breast Box
It is difficult to know how well Robert's arrangement for pouring the pulp onto the wire would have worked without actual trials. Donkin states that the lifting drum with its vanes like the paddles of a waterwheel did not deliver the pulp evenly to the shelf behind the delivery slot. The slot through which the pulp flowed seems to have had a fixed opening with no adjustments. The wire sloped from the point where the pulp was delivered up to the press rolls. Robert seems to have relied on the pulp running off the sides of the wire and back over the end (like the vatman tipping the surplus off his mould) to attain an even sheet of paper. As the machine is laid out in the drawings, there is only one driving handle and all the parts are constantly in gear. Therefore it would have been impossible to vary the speed of the wire without also varying the speed of the lifting drum. If the paper were becoming too thin, the lifting drum could not be speeded up by itself to deliver more pulp. Looking back with hindsight, we can see that this whole arrangement was modelled far too closely on hand papermaking techniques and skills. Improving the layout of the breast box, the slice to control the volume of pulp flowing out of the delivery box and the apron down which the pulp ran took a great deal of time and experiment over the next few years.

The Moving Wire
Robert realised that it was necessary to shake the moving wire on which the paper was made in order to 'close' the fibres and interlock them, as did the vatman with

the pulp on his mould. Therefore Robert supported the breast roller round which the wire passed on an adjustable frame. The roller was suspended from a metal cross-bar which was supported at each end by vertical bars. These were connected to a horizontal wooden beam below the vat. The whole frame could be raised or lowered by a single screw in the middle of the beam. It is doubtful if such an arrangement would have lifted the frame squarely. Two screws, one near either end, would have prevented the frame and the roller from tipping sideways, and so kept the wire in a better horizontal position.

This design fault may have led to one of the alterations in the 1803 patent. In order to give his paper the necessary shake, Robert had fixed an octagonal wheel on the main driving shaft so it would hit the frame and impart the shake. If the frame were not very stable, the shake would not have been very effective. In patent No. 2708, Gamble devised a positive action

> . . . by causing the pin to move in a groove or cavity, waving in the face of a
> wheel, or between two surfaces which shall confine the said pin both ways,
> and render it impossible that any jerk should be produced by the pin being
> driven off during the quick motion of the machine.

It is not clear how the shake was given to the wire in the 1807 patent, but the shake has remained a feature on most Fourdrinier machines up to the present day.

It is not known what type of wire was actually used, whether it was wove or laid. All the drawings indicate a type of laid, but as only three chain lines in a width of two feet are shown in the drawings of Robert's machine, this is clearly insufficient, assuming that the wire was made in the usual way. Since all the other parts of the machine are shown in minute detail, it would have been quite easy for the draughtsman to draw the chain lines. It is doubtful if even five chain lines (this allows for two more at the sides of the wire) could withstand the longitudinal tension to which the wire must have been subjected. A laid wire would have passed more flexibly round the small diameter of the rollers (only 2 in., 5 cm.,) but the twists in the fine chain wires would have rubbed against themselves and probably soon broken. In France, laid wire was still more common around 1800 than wove and it may be significant that the 1807 patent starts by saying,

> And accordingly we do describe the nature of our said machine . . . and do
> declare that the same consists in the use of a revolving web of wove wire. . . .

Might this indicate that there was a change to wove wire in England? Certainly in England the end rollers round which the wire passed were increased in diameter on subsequent machines which would have caused less fatigue on the wire.

Stretching from the breast roller to the press rollers were two horizontal bars placed either side of and just above the wire. Hooks attached at regular intervals along the sides of the wire could slide along these bars and keep the wire from sagging. The distance apart of these bars could be adjusted also to tension the wire across its width, but the wire itself was unsupported in the middle until it reached the press rollers.At first sight, this might be thought to give ideal drainage conditions with nothing to restrict the flow of water out of the pulp. In fact, it fails to recognise the dual contribution of the ribs in hand papermaking moulds. Not only do they keep the surface of the mould level and prevent it sagging when the paper is being made, but, through the surface tension of the water, they help to de-water the pulp by drawing the water down and along their length. Now, while a laid wire would have been less flexible in its width than a wove wire, it would still have needed the equivalent of the ribs underneath it to help drain out the water. With wove wire, ribs are vitally necessary for both support and drainage. It is possible that two wooden cross-bars of fir were suspended from the horizontal bars of the original machine but just how is not clear. In the 1803 patent, at the very end, we find another important addition,

> When the circulating piece is of considerable dimensions, I support it
> and prevent it from yielding downwards by one or more rollers revolving
> along with it.

Similar rollers, or more modern replacements, have been used ever since, both for support and drainage.

The tensioning of the wire across its width remained a problem for a long time. The hooks were replaced by buttons sliding in a slot in the 1803 patent, and with guide wheels in that of 1807. All these devices did, of course, help to guide the wire and keep it in the centre of the press rollers. They were made superfluous later by the supporting rollers and by the simple expedient of being able to alter the angle of one of the rollers underneath which deflected the wire.

The Formation of the Sheet

Robert hoped that he could pour the pulp out of the slot in his breast box onto the wire in a constant stream so that an even layer would be deposited on the wire as it moved away underneath. To vary the thickness of the sheet of paper, he tried to copy the vatman who tipped off some of his pulp from the mould. On his machine, the wire sloped up from the first roller to the press rollers so that the water would always run back to the lower end. By altering the inclination, more or less pulp would flow back over the first roller and into the vat. On subsequent machines, the wire has always been level and the thickness controlled by the amount poured onto it.

The width of the sheet was governed by the distance between the horizontal bars which guided the hooks. They acted as sort of deckles and, in addition, strips of eel-skin were sewn along the edges of the wire. This method was unsatisfactory because the only way of altering the width of the paper was by changing the wire and moving the positions of the horizontal bars. In the 1803 patent, the third improvement was to place on top of the wire what were in effect deckle straps which could be easily adjusted for width. These also have remained features on most machines until recently.

The wire passed between a pair of press rollers which were covered in felt. Here again, this was copying hand-making practices where the sheet of paper was couched from the mould onto one felt and then covered with another. Robert seems to have copied his wet press from the glazing rolls or calender. The main drive for the wire was through the lower roller but later this was moved to the final couch roller. The rollers could be screwed down to increase the pressure they exerted on the wire, yet, because they were so small in diameter, they may not have exerted enough force to de-water and consolidate the sheet sufficiently. Trials in Egypt on a replica machine built recently have confirmed that this is so and that the paper is very wet when it comes off the machine. The first sheet of drawings in the 1803 patent shows a second pair of press rollers, the dry press, mounted on their own frame which could be placed at the end of the actual paper machine. The paper was taken through these rollers between a pair of endless felts. This press was added in France, but no part of it was shown on Robert's original drawings. Although all the rollers on later machines were much larger and stronger than Robert's originals, it was still found necessary to pass the paper through a second dry press supported on a continuous felt after it had been taken off the wire, another feature which has been included on all subsequent Fourdrinier machines. Making the wire and the second dry press run synchronised proved to be very difficult for otherwise the wet web of paper would break or cockle. Donkin developed an expanding pulley for the driving belt so the speed could be adjusted easily.

Robert assumed that, once he had pressed his sheet, he could easily remove it from the wire, but at this stage he failed to understand what the coucher does with the hand mould to deposit the sheet of paper. The coucher puts one edge of the mould on the felt and rocks it across, lowering the other edge and raising the first. He also pushes down so that the wet paper is compressed. Water passes up through the wire covering and, when the pressure is released, flows down again, helping to wash the fibres off the surface of the mould. This happens on later papermaking machines, where the first press roll has become the couch roll at the end of the wire.

Robert put his single wet press two-thirds of the way along his wire and, with the driving force coming there, the last part would have run slackest until it passed round the roller at the far end. It was just in this position, where the wire was unsupported, that Robert put a wooden couch roller across the top on which the sheet of paper was to be wound. This was replaced very quickly by the second dry press mentioned earlier, but even so the layout was unsatisfactory. In 1803 or 4, a second wet press was added at the end of the wire in the position of later couch presses and the paper was transferred to a felt for a third pressing before being wound up on a reel.

Robert tensioned his wire by another screw vice on the final roller. This also had only a single screw, so that it would have had the same weaknesses as the frame supporting the first roller. With only one screw, it would have been difficult to keep the rollers parallel and, even though the hooks on the wire would have helped to guide it, a pair of screws, which could have properly adjusted the final roller, would have been a better solution.

The paper wound round what may be described as the wooden couching roller still had to be dried. There was no screw press capable of taking such a length, although the sheet was generally limited to twelve yards, or the size of a piece of wallpaper. The sheet had to be suspended on lines in the drying loft but its weight made it contract unequally as it was drying so that it was rendered useless for the

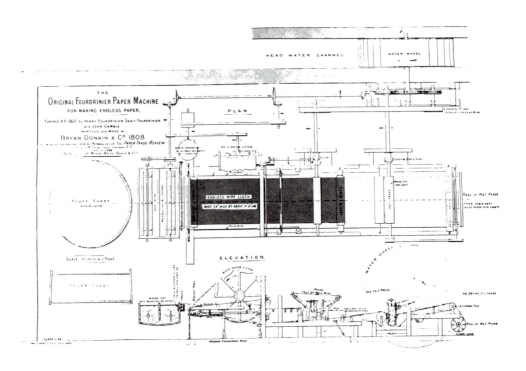

21. By 1808, Donkin had arranged the wet end of the Fourdrinier machine in a layout which changed little for the next 150 years.

wallpaper manufacturers. Later a better way of drying was evolved and many lengths were sold to Parisian wallpaper producers. The 1801 patent included a device that could cut the paper into sheets as it came off the dry press. By 1807, the paper was wound onto a reel. When the appropriate number of revolutions had been completed, the wad round the reel was cut through so what was one length became individual sheets. These, still in their wad, were dragged onto a cutting table where they were trimmed to the correct length and then taken away for drying. Whatever the weaknesses of Robert's machine, it did provide the stimulus to Gamble and the Fourdriniers, assisted by Bryan Donkin to develop it and turn it into a successful papermaking machine.

Later Developments

In June 1804, after spending some three months of trials on their second experimental machine, Bryan Donkin started to build a third which was installed at Two Waters Mill in 1805. This machine was 24 ft. 5 in. long and 5 ft. (7.4 x 1.5 m.) wide and was capable of making paper 54 in. wide. Although this machine had the advantage of being much better constructed through the experience gained on the earlier ones, much still remained to be done to make it work satisfactorily. By adding a short wire round his wet press above the machine wire,[8] Donkin was able to increase the speed of the machine from 20 ft./min. to 34 ft./min. These wires lasted about three months before they were worn out. Continued modifications and improvements added to the costs of all these machines so that up to June 1807, well over £5,000 had been spent on those at Frogmore. The Two Waters machine cost £4,203. 17s. 10 d. and a further £2,726 was spent on the mill itself and the drying house there. In addition, over £3,000 had been spent at Bermondsey on the engineering works.[9] Also in 1804 the Fourdriniers had purchased a mill at St. Neots in Huntingdonshire where they intended to set up John Gamble with two papermaking machines.

By 1807, the Fourdrinier machine, as it must now be called, had developed into a fairly satisfactory papermaking machine and promised to be a commercial success. In that year, Parliament granted an extension of the terms of the patents for fifteen years, although this was later challenged and reduced to seven.[10] In 1807, the St. Neots Mill was equipped with one machine for Gamble but the second was sold to J.B. Sullivan near Cork in Ireland for by this time the Fourdriniers were beginning to feel the financial strain. The St. Neots Mill cost £11,276 and its machine £2,754. Donkin was able to show that up to 1807, he had done work for the Fourdriniers amounting to nearly £32,000 of which £11,915 was attributable simply to experiments.[11]

Other machines were sold in 1807 to John Phipps of Dover, L. Smith, Peterculter at Aberdeen, E. Martindale of Cambridge and James Swann of Eynsham. While

Smith was granted his licence on 1 July 1807, it seems that the date for starting successful production was 1811, presumably because there was insufficient capacity at the Bermondsey works. The first use of machine-made paper by the *Aberdeen Journal* was on 26 August 1812.[12] Four more machines were sold in 1809 and the same number again in 1810. Donkin made nineteen machines between 1802 and 1812 but only fifteen in the next ten years. The prices of these machines ranged from £715 for one with a 30 in. (762 mm.) wire to £1,040 for a 54 in. (1371 mm.) one, depending upon size.[13]

The Fourdriniers claimed in 1805 that their machine was capable of doing the work of six vats in twelve hours. In 1806, when they were advertising their machine, they made a comparative estimate of the expense of attending seven vats and a machine running twelve hours per day. The cost of the hand mill was £2,604. 12 s. per annum but that of the machine only £734. 12 s., a saving of £1,870. When William Balston visited the Two Waters Mill in 1808, he was told that the machine would produce 84¾ reams (40,680 sheets) of 14 x 18 in. (355 x 457 mm.) paper in a twelve hour day, which would equal the output of eight vats according to the custom of Kent, where 10¾ reams per vat was accepted as a day's work.[14] The number of men employed was reduced from 41 to 9. The expense of making paper by hand at this time was 16 s. per hundredweight and by machine 3 s. 9 d.[15] However these figures did not include the capital cost of the other necessary processes such as beating, sorting or the supply of rags. In 1837, a papermachine with a width of 30 in. (762 mm.) was calculated to do the work of three or four vats and one with a 54 in. (1371 mm.) wire was equivalent to twelve vats.[16]

The Bermondsey works were leased to Donkin who actually built the machines and paid the Fourdriniers a royalty of between £80 and £200 depending upon the size. The Fourdriniers also ought to have received at the same time a royalty from £150 to £500 per annum depending upon the amount of paper produced by all users of the machines,[17] but many of them were refusing on petty technical grounds which led to law-suits to try and recover the money. Continual development and improvements at both the Two Waters and Frogmore mills cost more money. Leger Didot had been frequently in England, giving advice on new ideas, and he received £14,879 from the Fourdriniers between 1802 and 1807.[18] At the inquiry of 1807 into the case for the extension of their patents, the Fourdriniers were able to prove that they had withdrawn £60,228 from their stationery business to finance the invention. Although in 1809 their assets were valued at over £230,000, their balance sheet showed a loss approaching £6,000. On 8 November 1810, the Fourdriniers were declared bankrupt and in 1837 they claimed that they had made a net loss of £51,685 on their machine. Whatever the loss to the Fourdriniers, they had developed the machine on which most of the paper is produced in the world today.

VIII

RIVAL MACHINES FOR
MAKING PAPER

As long as there was little need for sizes of paper larger than could be made by hand, the driving force behind many people who tried to invent paper machines was the wish to produce paper with the least number of workmen and to use less skilled labour where possible. In France and in several parts of Germany, members of papermakers' unions had assumed dominant positions in mills which caused abuse, reluctance to work properly and strikes. Later, once the Fourdrinier machine was in production, there was another reason for trying to develop other types of machines, particularly those imitating handmade paper with individual moulds and not a continuous wire, because paper coming off the Fourdrinier machine had no watermark. Also in England there would have been initially the dues to be paid on patent royalties to the Fourdriniers and so other inventors hoped they would capture a profitable market if they could devise something similar.

On the request of the Académie des Sciences, Arts et Belles Lettres of Dijon, Phillippe Xavier Leschevin and Pierre-Joseph Antoine went to the small village of Poncy near Dijon in June 1811. There they visited the mould-maker Ferdinand Leistenschneider, who notwithstanding his German name, was a real Frenchman born in the Department of La Moselle. For about twenty years, Leistenschneider had been working on the invention of a machine for making 'endless' paper and the two visitors had been charged by the Académie to report on this. Leistenschneider's machine was a cylinder-mould of the 'dry' type, where the pulp was poured onto the outside of a drum covered with wire so the water strained through it. As the drum rotated, so the sheet of paper was taken off further round its circumference and passed through rollers for pressing. Since he was a man of limited means, he was unable to have a full-scale machine built, and could only show a simple model that apparently worked very well. The very favourable report helped the invention to get a patent for ten years on 19 November 1813.[1]

Meanwhile in England, Joseph Bramah had patented in 1805 a machine 'for making paper in endless sheets' that worked on a similar principle.[2] One of his

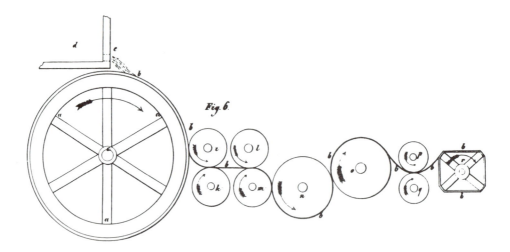

22. A diagram of Bramah's 'dry' cylinder mould machine, c. 1805.

suggestions was to pour the pulp onto the surface of what he called 'an universal revolving paper mould'. This was made in the same way as a hand mould with a cover of either wove or laid wire supported by ribs but turned into a cylinder on a framing like a waterwheel 'with a shield on each side of the upper extremity of its periphery, so made to fit the edges of the two extreme rings in a segment form nearly water-tight, as to prevent the lateral discharge of the fluid passing over the wheel'. The pulp was fed onto the wheel from a higher cistern and the water passed through the wire, leaving the sheet of paper on the surface. Opposite the feeding place on the wheel was placed a couching roller covered with felt and held in contact with the mould by springs. From here the paper was passed between two felt covered press rollers to have water squeezed out. If these did not remove enough water, there could be further sets of pressing rollers. What Bramah does not say is how the paper was conducted between his couching roller and the presses. There is no mention of any endless felt and, if there was none, it is doubtful if the sheet would have been strong enough to have supported itself and so would have broken frequently. His patent contains no drawings so we do not know how accurate is the one shown by W.A. Rust much later in his *The Technology of Paper* where the paper is taken off the couch roller by the press roller, an idea which would not have worked very well.[3] Once the paper had been pressed, it was taken over plates heated by charcoal or other fires or passed between the rollers of heated calenders 'so that when the paper leaves the machine it may be in a state fit for use'. This would appear to be the first time that heat was proposed for drying paper on any machine. It has been doubted whether Bramah actually ever made a machine on this principle. Much later this method was developed for board machines and so could have been made to work and is called the dry mould paper machine. It has proved to be unsuitable for making fine paper where the watermark is an important feature.[4]

At the same time, Bramah also patented a machine that imitated only the vatman's work by which sheets of much larger dimensions could be made. The pulp was contained in a storage chest where it was stirred by an agitator to keep it in suspension and then let out through a broad opening with a valve. Underneath this was a box in which the papermaking mould just fitted. The mould was covered by water in the bottom of the box and the correct quantity of pulp poured in from the storage chest. As the mould was raised, the flow from the storage chest was cut off and a valve in the box opened so the water drained out, leaving a sheet of paper on the mould. It is very doubtful whether this sheet would have been at all even with this method of pouring. When the full mould had been raised to its highest point, the operator could slide it out of the box sideways so it remained level. He replaced it with another and couched off the sheet manually 'and thus the process is continued with the utmost certainty and ease'.[5] Neither of Bramah's inventions seems ever to have been put into practice.

While the Fourdriniers were trying to develop their machine based on Robert's patent, they must have had misgivings about its success, for in 1806, Henry Fourdrinier obtained a patent for a method of making paper of indefinite length, both wove and laid, with separate moulds.[6] Possibly it was Didot who persuaded them to diversify their interests in this way for over the next fifteen years he was to take out further patents for the same idea. Unfortunately the 1806 patent contains no drawings nor any adequate description of the machine. The moulds, which passed over a series of rollers, were hooked onto each other as closely as possible to form one long mould so that a continuous sheet of paper would be formed as the pulp was poured onto them. The paper was taken off on a continuous felt upwards through a press while the moulds passed back underneath ready to be filled again. This machine never appears to have been used commercially, possibly because the paper was weak at the places where the moulds touched each other.

The next attempts at using a continuous chain of moulds were to make separate sheets of paper, taking advantage, like Henry Fourdrinier, of the fact that separate moulds could have watermarks. Between 1807 and 1812, Thomas Cobb, of Banbury in Oxfordshire, developed a quite complex machine with a chain of separate moulds.[7] It consisted of the usual stuff chest, breast box and delivery slice. Below this was a conveyor on which the moulds were placed. The conveyor was driven by hand so each mould remained under the pulp delivery long enough for it to be filled. On his first machine, the mould was taken off the conveyor and couched by hand but later the moulds were transferred to a second conveyor and passed under a roll round which a continuous felt travelled. The felt couched off the paper and passed it between two press-rolls from where the paper could be removed.[8] In spite of later improvements, this machine never achieved success.

1812 was also the year when Didot patented a machine made by Bryan Donkin for making paper on moulds in single sheets which was very ingenious and highly complicated. It reflected great credit on its maker, especially as he never really believed in it and always insisted that it would never turn out to be as great a success as the endless wire Fourdrinier machine. Its main advantage was that watermarked paper could be made on it but the problem of taking the sheets off the couching felts (a pair was used) without the assistance of a man was never solved. Possibly some of the most important features were the methods employed to wash and clean the felts. They were brushed by a brush roll, washed with a spray of water and then squeezed by a small press roll. Similar washing facilities were installed on Fourdrinier machines later.[9]

What was possibly the last attempt in Britain to invent a papermachine with separate moulds was patented in 1816 by the Scotsman, Robert Cameron.[10] The pulp flowed constantly onto the moulds as they passed underneath. That which missed the moulds was conveyed back to the storage chest by an Archimedes screw. The moulds were shaken to 'separate the water from the pulp, and makes it acquire a connected texture' (Patent). As the moulds began to incline downwards, the deckles were lifted off and the moulds passed to the couching section where they were inverted onto revolving felts or boards to which the paper was made to adhere. As the empty moulds passed back to their starting place, they were washed by a stream of water. The pieces of paper on their felts were passed 'betwixt the cylinders, where it is pressed and dried by steam' (Patent). This dried the sheets which were delivered from the machine on individual felt-covered boards. The most surprising item in the patent specification is these steam drying rolls which have been neither widely recognised nor used for some years. The answer may lie in the fact that the machine was probably not employed outside Cameron's own mill and possibly one other.[11]

There was a steady interest in the mechanical production of handmade paper on the Continent which achieved success many years later. The most important advocate of such a system was Max Sembritski, director of the Austrian papermill Schlöglmühl. In 1881, he acquired a German patent for a chain-mould machine which could produce paper of the highest quality.[12] It was built by Escher Wysse & Cie. at Zürich. Models were sent to Austria, Germany, France and in England to Turner, Symons & Co. at Tuckenhey Mill, Devon. At about the same time, French banknote paper began to be made on the Dupont machine, a vast contraption which had the forming part on one level of the mill and the drying cylinders on the floor above. It ceased to be used in about 1947.[13] In Holland, Walther Stegher designed a chain-mould machine in 1927 which was built by Werkspoor N.V. in Amsterdam for the Pannekoek Mill at Heelsum. Although the mill was destroyed by fire in 1944, this machine was repaired and set to work again at the mill of Van Houtum & Palm at Apeldoorn where it remained until it was broken up in 1963.[14] Therefore the principal of the chain-mould machine could be made to work but it was probably far too complicated for the early inventors.

John Dickinson, 1782–1869

After being educated at private schools, John Dickinson was apprenticed as a stationer to Messrs. Harrison and Richardson. It seems that he was influenced in his choice of profession by one of his intimate friends, Andrew Strahan, who at that time held the office of King's Printer.[15] Then in 1801 he commenced business on his own account and by 1804 was trading as a stationer at Walbrook in the City of London. His account books show that he bought paper from the Fourdriniers in 1805 and so was probably aware of their experiments. He obviously had an inventive turn of mind for in 1807 he patented a new kind of paper for making cartridges for cannon which would not continue to smoulder after the shot had been fired and so cause a premature explosion when the gun was reloaded.[16] He

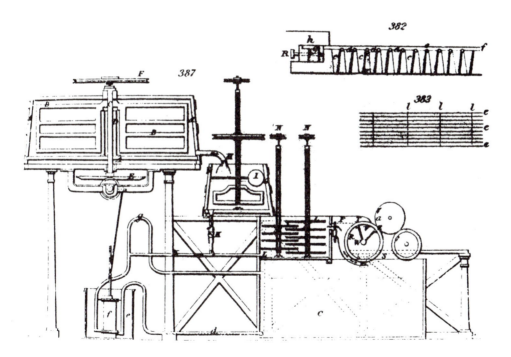

23. John Dickinson's first cylinder mould machine, 1809, had a complicated arrangement of storage chests and breastboxes with agitators to dilute and thoroughly mix the pulp before it passed into the vat and cylinder (R). Inside the cylinder is the vacuum trough.

mixed woollen rags or other material to the half-stuff so the paper was prevented from retaining sparks after the flame had gone out. Through his father's influence, it was adopted by the Board of Ordnance and proved its worth in the Peninsular Campaign and at Waterloo.

Earlier in 1807, Dickinson had patented a machine for cutting paper made on the reels of the Fourdrinier machine.[17] His account books from 1802 onwards show considerable and continuous payments to George Dodd, a well-known engineer of that time, which suggests that he was having experimental machinery made to his own specifications. Therefore it is not surprising to find a further patent taken out in 1809[18] for improvements to his cutting machine. It was this patent which also included his first papermaking machine.

For how long Dickinson had been attempting to make a papermachine is unknown, but he is said to have erected an experimental model on the roof of Andrew Strahan's office in Printer Street, London. It was evident that if he were to exploit the invention, he must have his own mill. Dickinson had been accustomed to buy paper from George Stafford of Apsley Mill, Hemel Hempstead, which he bought in 1809 with the financial help of George Longman. He was so successful with his machine that in 1811 he purchased the near-by Nash Mill. In 1815, he was insuring the buildings and machinery at Apsley for £15,500.[19]

The patent drawings of Dickinson's cylinder mould machine show considerable improvements over Robert's machine at the same stage of development. It would be interesting to know how much Dickinson copied from the experiments carried out by the Fourdriniers because his machine already had a storage chest with an agitator to keep the pulp continually stirred from where the pulp passed to a second chest, also with an agitator. Here the height of the stuff could be kept at a constant head through a ball-valve arrangement. Then the stuff flowed out through another pipe into which water was added to dilute it to the right consistency for making the paper which had to be considerably less, about four times thinner, than on a Fourdrinier machine. From this final chest, which again had agitators, the pulp flowed through pipes into the back or vat in which was placed the cylinder mould drum for making the paper. These pipes controlled the level of stuff in the vat which in turn controlled the thickness of the final sheet of paper.

The crucial part of Dickinson's invention lay in the drum or cylinder, partially immersed in the vat of stuff through which the water was strained, leaving the sheet of paper on its surface. Its construction was highly complicated and was described in his patent.

> I take a cylinder constructed so as to possess the following requisites; – In the first place it must be hollow, and open at the ends; secondly, the surface of the purphery must be like a sieve, with apurteues communicating with the internal part large enough to permit the passage of water, but calculated to intercept

the fibres of rag; thirdly, it must be so contrived that the surface will not yield from its perfectly cylindrical form, notwithstanding a very considerable degree of pressure upon it; fourthly, it must be furnished with broad flat rings for the purpose of covering part of its surface at the ends; there may be several pairs of these rings of different widths, in order to vary the proportion of the surface which is left uncovered, provided the same cylinder is intended for making different sized papers; fifthly, it must be hung upon an axis in a horizontal position, and firmly fixed in its bearings so that it may be turned by any convenient power; sixthly, the numerous small apertures on the external surface must open into a less number of larger ones, communicating with the internal surface, with solid interstices between them; seventhly, it ought not to be made of wood, because it would be liable to warp, nor of iron, because it would rust and injure the paper; brass or any other strong metal will be found most convenient.[20]

There were two reasons for making the cylinder hollow. The first was to enable the water to flow away through its ends. Dickinson used the pressure difference between the level of the stuff in the vat and that of the water inside the cylinder to pass the water through the cover of the cylinder and leave the fibres on its surface. At first this layer is extremely thin, but as the cylinder rotates, so more fibres are deposited and the sheet forms gradually at each part of the cylinder immersed in the stock.[21] Water will carry on straining through the cover until the cylinder rises above the level of the stock. The sheet of paper is taken off the cylinder near its top onto an endless felt. The various ways of regulating the level of the water inside the cylinder will be examined later.

Dickinson proposed covering the surface of his cylinder with two types of wire, one to produce wove paper and the other laid. The wove was made quite simply from an 'endless web of woven wire, which is drawn tight over it' (Patent). It was attached at the ends to rings slid over the main cylinder by plates and screws and then tensioned. A laid cover would be made in a similar way with the laid lines parallel to the main axle. Below either type of cover was a complicated construction which had to be perforated to allow the water through but yet had be strong enough not to deflect under the pressure of the couching roller. This is a problem which still faces manufacturers today. Dickinson used a solid brass cylinder across which he turned a small channel rather like a screw thread. At the bottom of this channel he drilled a series of small holes, tapering to a larger diameter at their lower ends inside the cylinder. Across this channel, he made a series of notches into which he let cross wires running in the opposite direction so that the surface of the cylinder resembled 'net-work with openings of an oblong shape'. On top of this he secured his cover. The purpose of this elaborate construction was not only to support the cover but to give it depth, like the backing wires of a hand mould, through which the water could drain more quickly. The cover had to be drawn tight round the cylinder to prevent it working loose and sagging.

Dickinson soon hit one of the problems of the cylinder mould machine which is that, when it is desired to make sheets of paper, the diameter of the cylinder, and its width, must be calculated to give either that sheet size or multiples of it. While it is possible to blank off sections, this wastes the capacity of the machine. Dickinson tried to divide his cylinder into sections, but, on his first machine, he was unable to use cylinders of different diameters through the way he mounted them, as has been possible on later ones.

In his first patent, Dickinson did not construct his vat so that it fitted equally up either side of his cylinder. It was lower on the side where the empty part of the cylinder entered. This meant that he had to put some sort of sealing strip, probably leather or woollen cloth, across the lip to prevent the stuff running out. Then because the ends of his cylinder were hollow, he allowed the water to run away at the bottom of either side, and the patent drawing shows no method of controlling the level within the cylinder. This would have given him the full head of the difference in height between the level of the stuff in the vat and the bottom of the cylinder. But it meant that he had to seal the outer circumference of the cylinder against the sides of the vat. Now with the cylinder held rigidly in bearings, it must have been very difficult to make the seal effective, particularly with the materials available in those days, and this was changed in Dickinson's next patent. The water, after it had passed through the cylinder, was collected in a cistern and recycled by being pumped back for diluting the stock. The flow of water through the cylinder and the consistency of the stuff had to be so arranged that the correct thickness of pulp had been deposited on the surface of the cylinder at the point where it emerged from the vat.

The reason for making his cylinder hollow, without any supporting ribs, was to enable him to fit a vacuum trough inside it to draw out the water from the sheet of pulp or paper. This appears to be the first recorded use of this technique in the history of papermaking and was applied to the cylinder mould machine long before the Fourdrinier. Dickinson fitted a 'V' shaped or triangular trough under the upper surface of the cylinder. Its bottom was about level with the centre of the cylinder so it could be supported by a pipe running through the axle bearing. The top rubbed against the surface of the cylinder where there was another leather seal. When the whole machine was started, a pair of vacuum pumps was set in motion which created a partial vacuum in the trough. As the top of the trough was above the level of the stuff in the vat, atmospheric pressure would push on the newly formed sheet of paper as it passed, still on the surface of the cylinder, over the vacuum box so that water would be forced through the cover and into the box from where it ran out by means of the central supporting pipe.

By the time the circumference of the cylinder had reached the further edge of the trough, the sheet of paper was sufficiently compacted on it so that it suffered no disturbance when it was pressed by the couching roller. The point of contact of the couching roller was just above the far edge of the trough so that any water that was pressed out here was drawn away by the vacuum. This roller had a smooth surface so that the paper would adhere to it and be lifted off the cylinder. The paper travelled around this roller until it was pressed by a second roller which had a 'puvious [sic] surface, consequently the paper will be produc'd [sic] sufficiently dry for leading off to the cutting'. The couching roller could be weighted to increase its pressure but probably these pressing arrangements were found to be inadequate because they were altered in the next patent. We are not told how the paper was reeled up nor taken away from the machine and dried. Machines could be built without the vacuum trough, in which case Dickinson would support the surface of the cylinder with strutts inside it.

One of the complaints later made against the cylinder mould machine was that paper formed on it could be easily torn down its length. Dickinson specifically denied this in his patent and claimed that,

> in making paper by the machinery above described the stuff is perfectly
> well shut without any shaking, the fibres of rag being deposited gradually in
> a longitudinal direction by means of the friction which takes place upon the
> cylinder, in consequence of its motion being in an opposite direction to that
> of the stream of pulp, the effect of which is to smooth down the fibres of rag
> as they are laid upon the cylinder, and it is necessarily continued during the
> whole time of the formation of the paper, and must be uniform throughout
> every part of it. The reason of introducing so large a quantity of water into
> the pulp is in order that every fibre may be afloat separate and at liberty to
> take a direction according to the influence of these causes.[22]

Dickinson envisaged a cylinder diameter of fifteen inches which would form paper equal to a sheet size of 22 x 17½ in. (56 x 44.5 cm.), weighing 20 lbs. per ream.

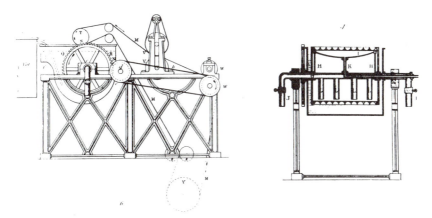

24. Dickinson's second machine, 1811, was considerably modified with better pressing arrangements and facilities for washing the felt. The ends of the cylinder were enclosed and the water syphoned out through pipes.

The rate of production was about 36 ft. per minute. The thickness of the paper could be changed either by adjusting the level of pulp in the vat, or by varying its consistency, or by altering the speed of the machine.

During the next couple of years, Dickinson made considerable modifications which he patented in 1811.[23] The result was a machine which more closely resembles many of those running today. First of all he extended the vat so that the cylinder was immersed equally on both sides. This gave more time for the fibres to be deposited and saved having to seal the lip where the cylinder entered the vat. This reduced friction and enabled the machine to be speeded up so that it could run at 48 ft. per minute.[24] Then he made the ends of his cylinder solid and again reduced the necessity for complicated seals around this part because he needed only to stop the stuff passing out around the axle. He drew off his water through pipes contained within the hollow axles of the cylinder. Inside the cylinder, these water pipes almost touched the bottom and, by extending them outside below the level of the lower surface of the cylinder, he could extract the water by syphoning it. Valves controlled the rate of flow so he could adjust accurately the water level inside the cylinder and so the rate of formation of the paper. He retained the vacuum trough which again acted with pipes passing through the journals of the main axle. The method of construction of the cylinder itself was slightly simplified to give freer passage for the water while yet retaining its strength.

Instead of a couching roller with a smooth hard surface, he covered it with several layers of woollen cloth to make the surface moderately soft. Round it passed an endless felt which actually picked up the paper off the wire on the cylinder. Again this couching cylinder could be weighted to help press out the water. From the couching roller, the endless felt passed through two roller presses to squeeze out more water when the sheet could be reeled up ready for taking away for cutting and drying. The reason for two presses was that the same amount of water could be taken out with gentler pressure than if only one were employed. Round the top roller of the first press it was necessary to introduce another short endless felt to prevent the paper from sticking to the surface. After the paper was taken off the main felt, this passed to the bottom of the machine where it was kept taught and washed by water being poured on it which was squeezed out by another pair of press rollers. Finally the vat could be sectioned so that different colours of pulp could be fed into them which would produce paper with different colours or stripes in the sheet.

Possibly Dickinson was removing too much water with his vacuum box for he began to have difficulty in lifting the web of paper off the surface of the cylinder. So he arranged that the air which was sucked out of the vacuum box would be recirculated and blown through the surface of the cylinder from inside just beyond the point where the couch roller pressed in order to blow the paper off the wire. He also arranged for a jet preferably of clear spring water to wash and clean the surface of the cylinder. There was an internal trough to draw off and carry away this water to prevent it mixing with the pulp.

By this last patent in 1814, John Dickinson had developed a machine that produced paper of high quality. The advantage which his machine had over the contemporary Fourdrinier lay in the comparatively small degree of wire-mark in the paper. One result was that the surfaces of both sides of his paper were nearly the same and his paper had a satin-like character derived from the fibres being laid in the direction in which the cylinder revolved in spite of what had been claimed in the patent. In 1812, Samuel Bagster had embarked on the publication of his *Pocket Reference Bible* and Dickinson had succeeded in producing a thin, trough, opaque paper which exactly met his requirements. The New Testament was issued in 1812 and the Old in the following year. Their immense success resulted in a large sale of paper from Apsley and Nash Mills.[25] Five years later, it was commented that

> The firm of Longman Dickinson & Co. has recently improved even upon the excellent Didot's (Fourdrinier) machinery and has produced a sort of India-paper-tinted 'article', quite delightful in colour, and apparently of equal excellence in substance. There is a story extant that a few of the principal London manufacturers of paper made a *bet* respecting the production of the finest 'article' in the trade; and who should *win* this bet but the House of Longman Dickinson & Co.[26]

By 1817, Dickinson had improved other parts of his machine. He anticipated board machines when he made two-ply paper by suspending a reel of wet paper over the couching felt and drawing off one sheet of thin paper from this reel onto the wet sheet picked up on the felt. The two were combined together in the wet press. A paper for printing from copper-plates was made in this way with a thin soft absorbent surface placed on top of a much stronger backing paper. This process was so successful that French paper ceased to be imported for this purpose.[27] At the same time, Dickinson also developed machines for pasting two or four lengths of paper together to make boards.

John Dickinson's most important development at this period was the introduction of drying cylinders heated by steam. At first he used only one cylinder about four feet in diameter situated between the second and third sets of press rolls. By the time the paper had passed round the hot cylinder, so much water was evaporated that the sheet was nearly dry, so dry that no humidity was perceptible to the touch. More than one steam cylinder could be used in series but then the thickness of the cast iron shell of each cylinder would be increased so that the heat would be applied moderately.[28] Whether the outsides of Dickinson's cylinders were smoothed and polished is doubtful because he recommended against boring the inside to attain an even thickness in the walls for fear of opening up blow-holes in the metal. Another problem which he faced was cockling of the paper. While the paper was supported on endless felts through each of three presses, none of these passed round the cylinder and so the paper was free to contract and cockle as it was drying. He recommended that the speed of the machine should not exceed fifteen feet per minute. This was the first application of steam-heated drying cylinders to a machine making endless sheets and precedes the patent of T.B. Crompton in 1820.

The drawing in Crompton's patent shows a line of six cylinders. The invention he claimed

> consists in conducting the paper by means of cloth or cloths against the heated
> cylinders, which cloth may be made of any suitable material; but I prefer it to
> be made of linen warp and woollen weft.[29]

Each cylinder was fitted with its own felt, a feature which continued until at least 1830. The felts prevented the paper from cockling while it was drying and overcame the problem Dickinson faced. However, in Crompton's arrangement, each cylinder would have needed to have its own guiding and tensioning system for its felt. One disadvantage was that the felts on the first cylinders could never have been dried by passing over later ones so would have remained wet. Crompton also envisaged heating his cylinders by steam. At the end of his drying cylinders, Crompton placed a cutter to chop the paper into sheets. When combined with Crompton's drying cylinders, the Fourdrinier machine became feasible for large production runs.[30]

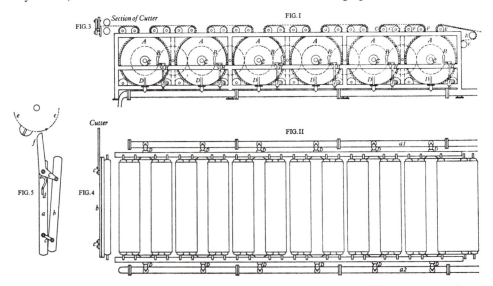

25. In 1820, T.B. Crompton arranged his drying cylinders in a single row with a separate felt for each one.

Making these cylinders was a skilled job. To cast such a cylinder required the right metal and the correct proportioning of the sections to enable it to withstand changes in both temperature and pressure while yet retaining a perfectly smooth regular surface free from pits and pinholes in which the paper might catch.[31] The surface had to be turned in a lathe to give a smooth polish. Crompton used felts round his cylinders while Dickinson did not, which may have been one reason why Dickinson's lead was not followed by other manufacturers. Then Crompton's invention was more usually applied to Fourdrinier machines but in any case it would appear that Dickinson did not seek to exploit his patents by selling rights to others. In an account of the Dickinson cylinder mould machine which appeared in the *London Encyclopaedia* of 1829, it was stated that

> The formation of the [papermaking] cylinder is very elaborate and the apparatus essential to the process so complicated that the machine has not been adopted by any other manufacturer except the original inventor, Mr. Dickinson: but in his management it works with great precision and effect, and produces paper of very good quality.[32]

So it would appear that no one else in England had adopted the cylinder mould machine up to that time.

In fact the writer of the above comment was wrong because by that date the cylinder mould machine was firmly established in America. In August 1817, at his papermill on the Brandywine Creek two miles above Wilmington, Delaware, Thomas Gilpin had started up the first paper machine to operate in America.[33] It was a cylinder mould machine. Now Dickinson placed greater reliance on secrecy than on patents, and it was reported that 'all his machinery was made at his own shops under the same rigid secrecy'.[34] One story is that one of his workmen, Lawrence Greatrake, left England and went to America, taking his knowledge of Dickinson's machine with him. Another is that Greatacre, the Gilpins' papermaker, visited England and stayed three days with the Dickinsons.[35] With Joshua and Thomas Gilpin's undoubted mechanical genius, they developed, together with Greatrake, a viable papermaking machine on which they soon produced paper noted for its fine uniformity of texture and freshness of colour. The machine delivered paper 30 in. wide (76.2 cm.) and, attended by two men and a boy, made as much as twelve men and six boys formerly.[36] Whatever the truth about the way the secret reached America, Dickinson always felt very bitter about Gilpin, whom he accused of having 'bribed Greatrake, his right-hand man and took him to America, forestalling him there'.[37]

The first Fourdrinier machine was not running in America until 1827[38] and to begin with the cylinder mould machine was more popular over there. It was probably in America that the cylinder mould machine began to be used extensively for making board as well as other papers. In 1829, there were two American patents for obviating the defect in paper made upon these machines for the paper could be easily torn in the direction in which the fibres were laid by the rotation of the cylinder. Isaac Saunderson introduced a horizontal fan stirring the vat while Reuben Fairchild tried a vibrating cradle of metal. Saunderson also applied sheet forming rollers so he could make press papers, pasteboard and bandbox paper.[39] In the following year, 1830, Dickinson himself took a step towards developing the modern board machine when he introduced two vats from which sheets of paper could be drawn simultaneously and combined into a board as they passed together through the press rolls.

In the meantime, there were advances in creating watermarked and security papers on the cylinder mould machine. On the Continent, Leistenschneider had been busy developing a second type of machine which has been considered to be an adaptation of the cylinder mould machine, producing not an endless web but separate sheets which were automatically gathered up into half reams. In the Royal Library at the Hague in Holland, is a sheet of wove paper 8¾ x 15¾ in. (22 x 40 cm.) with a single line watermark in both halves of the sheet. In the left half is a portrait of the Emperor within a circle, the head crowned with a laurel leaf, and surrounded by the inscription, 'NAPOLEON EMPEREUR ET ROI'. At the lower edge of the left half of the sheet we find the words, 'FAIT PAR FERDINAND AN 1813'.[40] This could be the earliest watermarked paper made on any machine.

In the early 1820s, John Dickinson was trying to make laid paper again, but his more important development for security papers was patented in 1829. This was for

> A method of introducing into paper, cotton, flax or silken thread, web, or lace,
> or other material . . . so that the thread, lace, or other material may constitute
> the internal part of the paper.[41]

The material that was to be incorporated into the paper was wound onto bobbins which were suspended above the vat. The ends of the material passed over a guide roller so they were in correct alignment before they descended into the stuff on the side where the cylinder rotated into the pulp. Here they were drawn onto the surface of the cylinder part of the way round where a thin layer of pulp had already covered the wire. As the cylinder continued to rotate, now taking the material round with it, the fibres continued to be deposited on the surface, covering the material so it was incorporated into the middle of the paper. To begin with, the bobbins might have to be turned by hand to ensure that the material or threads were drawn round with the cylinder, but once they had started, they continued by themselves. This is the way the metal threads are inserted into our bank notes today. Because the security wires, or tapes with magnetic codes on them, are actually incorporated into the paper as it is being formed, it is very difficult to forge. This was another important contribution by John Dickinson to papermaking.

THE NINETEENTH CENTURY: DEMAND OUTSTRIPS RAG SUPPLIES

Rags are as beauties, which concealed lie,
But when in paper how it charms the eye:
Pray save your rags, new beauties to discover,
For paper, truly, every one's a lover.
By the pen and press such knowledge is displayed,
As wouldn't exist if paper was not made.
Wisdom of things, mysterious, divine,
Illustriously doth on paper shine.
Boston News Letter, 1769

During the nineteenth century, the pace of change gathered momentum as the Industrial Revolution affected more and more industries and created new ones. The demand for paper continued to rise, partly because an expanding commerce in itself increased consumption and partly because new uses were found for different types of paper. This can be illustrated in two areas of activity, in printing and in the postal services. Printing is commonly considered mainly in terms of the newspaper, the periodical and the book, but there has always been a third class of work, the jobbing printer producing local advertisements, tickets, and small jobs for private people. It is estimated that in 1851 there were 563 newspapers including 17 dailies which had risen in 1867 to 1,294 with 84 dailies. Late Victorian Britain had about 2,000 newspaper establishments and its book-printing firms were confined to two dozen towns but jobbing printers numbered at least 8,000.[1]

In 1800, the wooden hand press reigned supreme, but about then, Charles, the Third Earl of Stanhope, designed a completely new press made of iron with an ingenious compound lever action. This permitted not only printing the whole of the forme of type at once, but at the same time allowed formes to be enlarged. By his invention, not only was the maximum size of forme increased from four octavo pages to sixteen but the speed of printing rose to around 300 impressions per hour. Other iron presses appeared later, with George Clymer's popular 'Columbian' press being introduced to England from America in 1817. In 1822, R.W. Cope invented his 'Albion' press. This was followed in the next few years by many other designs which by 1850 had taken the development of the hand press as far as it could go.

On a hand press, letterheads, business cards, circulars, etc., which were the mainstay of the jobbing printer, could be printed only with difficulty. The answer to his problems came with the appearance of the hand-fed treadle platen, another American invention. The first type to be marketed in this country was H.S. Cropper's 'Minerva' platen which was made in Nottingham. The first of these were available by 1860 but they did not find general acceptance until the 1870s. Other manufacturers followed with their own versions, for example the popular 'Arab' and the later 'Adana'. With automatic inking, capacity was raised to 1,000 impressions per hour but the paper was still fed by hand. Towards the end of the century, even this speed was proving too slow for the ever increasing demand for printing so by 1890 there had been developed platen presses driven by power with automatic feeds. These reached speeds in the region of 4,000 impressions per hour. There was little further improvement in automatic platens until the 'Heidelberg' appeared around 1926, originating in Germany. Its improved feeding mechanism enabled speeds of around 5,000 impressions per hour to be reached.[2]

By the 1850s, new, simple types of cylinder flat-bed presses, such as the 'Wharfedale' had been invented and were widely adopted by book printers. In 1854 Thomas Main and William Conisbee in England introduced a flat-bed press with a speed of 1,200 to 1,300 impressions an hour while in America, at the same time, Hoe patented his 'little Astonisher' which had a rate of 3,000 to 4,000 impressions. The reason for this name was that it surprised printers for it could do fine work which they had thought only the hand press was capable of performing.[3] These early presses remained hand-fed, but as the century progressed, speeds were increased and ultimately automatic feeding was introduced. All these presses printed on individual sheets of paper.

All the presses mentioned so far used type mounted on a flat bed. The development of really high-speed printing could be achieved only with the type itself placed on a revolving cylinder. On 29 November 1814, *The Times* was printed on a steam-driven rotary cylinder press at the rate of 1,100 impression an hour, using the invention of Friedrich Koenig.[4] In order to try and reduce the amount of paper that was being printed, in 1818, the Government ordained that the dimensions of any newspaper should not exceed 22 by 32 in. (560 x 815 mm.). For many years, *The Times* continued to lead in the printing world. A circulation figure of around 2,000 in 1790 had increased to 10,000 in the 1830s and 65,000 in 1861.[5] By 1836, it was being printed on a press built by Applegarth and Cowper at the rate of 4,200 impressions.[6] In 1850, this had been increased to 10,000 on an improved Applegarth machine which was superseded by an American one by Hoe. While this could provide 20,000 impressions an hour, the paper was still fed in by hand from five different levels and twenty-five men and boys were required to tend it.[7]

Possibly this should have satisfied Tomlinson's complaint when he wrote soon after 1850,

> The increased supply which they afforded stimulated an increased demand; nor need we feel surprised that a journal conducted with so much intelligence and honesty of purpose, written with so much skill – often, indeed, amounting to genius, – that a journal which has long proved itself to be the consistent advocate of the poor, the injured, and the oppressed, and the unflinching opponent of the selfish, the unjust and the mean; that such a journal, shedding not only an informing light on the politics and news of the day, but advocating with a master-hand the claims of literature and science, the fine arts, and the useful arts, – of everything, in short, that is calculated to advance civilization, and to promote the dignity and happiness of the human race, – we cannot wonder, we repeat, that the circulation of such a paper should be to a great extent limited only by the power of multiplying the number of its copies.[8]

What Tomlinson did not say was that by 1850, there were beginning to be serious difficulties in finding enough rags from which to make the paper. The paper for *The Times* alone, 'for a week's impression fills 13 waggons; and that the whole of one day's impression, taking at 42,750 copies, weighs about $7^1/_2$ tons'.[9] At about the same time, the *Illustrated London News*, on one occasion, sent forth no less than half a million double numbers, or one million sheets exceeding 70 tons in weight.[10] Yet between 1815 and 1861, the price of *The Times* fell from from 7 d. to 3 d. per copy and that of the *Manchester Guardian* from 7 d. in 1821 to 1 d. in 1857.

Production of even more copies at a faster, cheaper rate was achieved through developments in three different spheres. Until 1855 when the Newspaper Tax was repealed, the Excise had insisted on stamping every sheet of newspaper individually so there was little point in printing continuously. In 1865, a device, which first came into use in Philadelphia and which was patented in the following years by *The Times*, enabled a continuous roll of paper to be fed in. This has changed the course of printing and the character of the industry supplying the paper.

As long as the type had to be set up by the time-consuming hand process, there was little gain in using more presses to print more copies. While stereotyping, that is making a mould from a bed of type and casting identical copies of it, originated around 1800, it was not until the 1860s that this was adapted in America to rotary presses.[11] Then a sheet of wet paper board, the flong, was pressed onto the face of the type so a reverse image was formed in it. The board was dried in a mould curved to the outer circumference of the printing cylinder. The curved flong was then used as one side of a casting box into which type metal was poured to create part of the surface of a cylinder which could be fitted into the printing press. To make the paper required for the flong became a separate part of the paper industry, but production ceased in England during 1986.

High speed printing would have been impossible unless the setting of type kept pace as well. It was a slow process to pick each letter out of its space in the tray and arrange it, word by word, line by line, in a composing stick. By 1842, Henry Bessemer, later the inventor of the convertor for making iron into steel, had produced a machine called the 'Pianotype' which achieved a speed of 6,000 letters or spaces an hour. The variety of typesetting inventions soon became innumerable, but it was not until the early 1890s that the two machines were invented, which became the standard ones until recent developments in photographic and computerised methods. Both were American ideas, with the 'Linotype', which as its name implies, cast a line of type, becoming more popular for newspaper and periodical work, and the 'Monotype', which cast a letter at a time so corrections could be made more easily, being used more by the book trade.[12]

The effect of all these inventions was to reduce dramatically the cost of printing. In 1837, when giving evidence before the Parliamentary Select Committee to see whether Henry Fourdrinier should receive some compensation for his invention, George Clowes, the famous printer, stated that 'The Fourdrinier machine had been very beneficial to the printing trade, materially reducing the price of paper and enabling them to produce books at a much cheaper price'.[13] Equally this could not have been achieved without a corresponding improvement in printing techniques for Clowes said that whereas formerly they used to go to press with an edition of 500 copies, they now printed 5,000. Tomlinson commented about the cheap literature of his day in 1850,

> That which the railway, the locomotive engine, and the steam-boat have
> done for locomotion and traffic, the printing-machine has accomplished for
> education and intelligence. Without the printing-machine, cheap literature
> could not have existed; English Bible Societies could not have scattered
> in such prolific abundance the word of God over the world, and the
> newspaper-press – that true exponent of a free people – must have remained
> comparatively ineffective.[14]

Improved communications, particularly through the spread of the railways after 1830, stimulated the demand for paper through quicker, better postal services. A uniform rate of two pennies a letter had been instituted for London in 1801, but outside the capital there were very heavy charges. It cost 8 d. to send a letter from London to Brighton and 1 s. 1½ d. from London to Edinburgh. The rate was based on mileage and also per sheet of paper so most letters consisted of a single sheet, folded with the message inside and the address written on the outside. Rowland Hill suggested that there should be a uniform postal rate based on weight. Postage would be paid by the sender and not the recipient and he proposed two ways of showing that the charge had been met. Either there could be

> . . . stamped covers or sheets of paper . . . supplied to the public from the
> Stamp Office or Post Office . . . and sold at such a price as to include the

postage . . . Covers, at various prices, would be required for packets of various weights; and each stamp should have the weight it is entitled to carry legibly printed with the stamp,

or there could be 'a bit of paper just large enough to bear the stamp, and covered at the back with a glutinous wash', to be stuck on letters.[15] After the introduction of the penny post in 1840, the number of letters carried within the United Kingdom rose from about 75,000,000 to 196,000,000 in 1842 and the gross revenue to about 63 per cent of the former receipts. By 1849, nearly 329,000,000 letters were posted.[16]

The papermakers gained from this sudden increase in three ways. First there was a greater demand for ordinary writing paper. Then there was a demand for paper for the stamps and, finally, there was a new demand for envelopes because the weight limit allowed the sheet of note paper to be enclosed within some form of protective wrapper. Envelopes, or 'pockets' as they were first called, were coming into fashion by 1835. On 13 March 1835, Fanny Dickinson wrote to her brother at Eton 'on blue notepaper enclosed in a *pocket*' and on 6 April 1836 she was busy 'making envelopes all day' to send out invitations to a party.[17]

By 1850, it was realised that envelopes needed to be made from a specially strong kind of paper to withstand any wetting they might receive from rain and so specially glazed paper was produced. At first the relatively small numbers of envelopes required had been made by being cut with a simple guillotine and sold flat, to be folded and gummed by the retailer or purchaser. By 1845, Rowland Hill's brother Edwin and Warren de la Rue had developed an envelope machine far enough for it to be patented.[18] With two operatives, it could produce some eighteen or nineteen thousand ungummed envelopes a day and by 1850 it was making gummed ones. These machines were improved quickly so that by June 1887, at Apsley Mill, Dickinson's were boxing three million envelopes a week, of which some 5 per cent were hand made.[19] On the latest machines, the paper was cut, folded, glued, the corners turned over and stuck, and the flap held open long enough for the glue to dry on that when it too was folded over. Then the finished envelopes were placed in a long trough with every twenty-fifth one a little to one side so that all the girl had to do was to put a paper band round each batch. Even the banding was done automatically on some machines.

Rowland Hill had proposed two methods for showing that the postal charge had been made and so the Government held a competition to design a scheme. Out of more than 2,500 entries, William Mulready won by recommending an envelope, to be sold flat and ungummed, with an elaborate engraved design to prevent forgery. Eventually a trial was made of 'Mulready' envelopes which were printed on paper supplied by John Dickinson and placed on sale on 6 May 1840. This paper was made on the cylinder mould machine with the security threads running through it. Dickinson had spent a long time perfecting this technique and managed to have it accepted in place of watermarked paper. At the same time, it was decided to try stamps which could be stuck onto envelopes, and so the famous 'penny blacks' were issued. These were printed experimentally on Dickinson thread paper, as were later the Penny Red and Twopenny Blue, and the embossed 1 s. and 10 d. octagonal stamps.[20] One reason that the use of Dickinson's special paper was soon discontinued, and hence the abandonment of the Mulready envelopes, was the fear of his obtaining a monopoly position.

In 1842, William Egley, an English artist, produced the original Christmas card and started a fashion which has become immensely popular ever since. This was followed a little later by Valentine cards. Not only did this encourage the development of different types of fancy paper, with for example lace patterns for the trimming around Valentine cards, but prodigious quantities of paper and card were consumed in their production. Then in 1870, the Post Office introduced post cards and gave de la Rue's a contract for supplying an hundred million cards, each stamped with an halfpenny stamp. This was, of course, regarded by other manufacturers as a monopoly and the business was opened to the paper industry. In 1883, nearly 160,000,000 cards were sent through the post, showing the great increase in the demand, and supply, of card for this new form of communication.

During the early part of the nineteenth century, the British paper industry remained in a sheltered position from foreign competitors through the continued import taxes. To begin with, the level of import duties on paper, as on most other commodities, was dictated by the needs of war. The import of rags began to be taxed in 1803 at a rate of around 25 per cent, which was about the same as that on paper itself. Both taxes were raised in 1809 to a net increase of around 33⅓ per cent on the old rate and in 1813 there was a further temporary increase, giving a total increase of around 58 per cent in seventeen years. The import duties were substantially reduced in 1825 and again in 1842. The duties on rags were abolished in 1845 but they survived on paper until they were finally abolished in 1861.

Import Duties on Paper, Board and Rags, 1802 – 1861

Year	PAPER		BOARD	RAGS
	Class I *per lb.*	*Class II* *per lb.*	*per cwt.*	*per ton.*
1802	1 s. ½ d.	6½ d.	£2. 2. 0	15 s. 9 d.
1809	1 s. 4 d.	8 d.	2. 17. 4	£1. 1. 8 in British ships
1819	1 s. 7 d.	10 d.	3. 8. 2	1. 6. 0 in British ships
1825	9 d.	3 d.	3. 8. 2	5 s
1842	4½ d. +5%	3 d. +5%	1.10.0+5%	6 d + 5%
1853	2½ d.	2½ d.	N.K.	abolished 1845 [21]

124

The rate of the Excise Duties followed a similar course, being gradually reduced as the century progressed.

Excise on Paper and Board, 1802 – 61

Year	PAPER Class I per lb.	Class II per lb.	BOARD per cwt.
1802	3 d.	1½ d.	£1.1.
1803	3 d.	1½ d.	£1.1.
1836	1½ d. for all classes		
1840	1½ d. = 5% for all classes.[22]		

Agitation against the Excise duties continued throughout the first half of the century by radicals who wanted Free Trade. A flat rate of duty on all papers produced a heavy imposition on coarser grades with the tax varying from 22 per cent on the finest sorts to about 200 per cent on the coarsest.[23] The first objectives of these agitators succeeded when the Advertisement Tax was abolished in 1853 to be followed by the Newspaper Stamp Tax in 1855.[24] This left standing alone the paper duty, which in 1850 had produced about twice as much revenue as the combined yield of the other two.

The tax on paper was seen as a tax on knowledge because it affected books and other printed materials, although a large part of the revenue came from other types of paper such as wall hangings. The problem of the paper duties came to the fore in 1860 over the trade treaty with France when Gladstone proposed to abolish both the Custom and the Excise duties on paper. This would open the market to foreign competition which the papermakers claimed would be unfair because in some countries the paper industry was sheltered by taxes on the export of rags. Therefore the British papermakers found themselves in the strange position of fighting against this legislation which included the repeal of the Excise duties, something they had wanted for years! Gladstone introduced the Paper Duty Repeal Bill in February 1860 which passed its second reading with a majority of 53 in March 1860 but got through its third reading by a bare 9 votes on 8 May, only to be rejected by the House of Lords on the 21st of that month. This precipitated something of a constitutional crisis and the paper industry found itself at the centre of general public discussion. However, the Radicals and the printers and others who were fighting against the taxes on knowledge, finally won the day. First, the Customs duty was reduced to equal that of the Excise during August 1860 and, in the budget of 1861, Gladstone introduced the repeal of all taxes on paper to take effect from that October. In fact, the paper industry was not ruined by the removal of the protective barriers and events were to show that demand increased for a greater part of the rest of the century.

Up to about 1860, the main source of raw material for papermaking was rags. Papermakers complained about the shortage of rags for centuries, but during the first half of the nineteenth century, rising production of paper in Britain created severe problems in the supply of this vital commodity. Between 1715 to 1855, production in England rose steadily from 2,500 tons to over 100,000 tons per annum. This increase was due partly to the growth of population, but the consumption per head was expanding also. From around 2.5 lb. in 1800, it had risen to about 8 lb.

Paper Production in the United Kingdom

YEAR	PAPER PRODUCED (LBS.)
1820	44,539,509
1821	48,204,927
1822	51,312,606
1823	54,214,611
1824	57,760,949
1825	62,106,933
1826	49,744,993
1827	60,040,287
1828	64,544,803
1829	59,404,709
1830	63,686,802
1831	62,738,000
1834	70,988,131
1835	70,655,287
1836	82,145,287
1841	97,103,548
1844	109,495,148
1845	124,247,071
1846	127,442,482
1847	121,965,315
1848	121,820,229
1849	132,132,660
1850	141,032,474
1851	150,903,543
1852	154,469,211
1853	177,633,010
1854	177,896,224
1855	166,776,394
1856	187,716,575
1857	191,721,620
1858	192,847,825
1859	217,827,197[26]

in 1860.[25] Such an increase would have been impossible without the introduction of the papermaking machines.

The paper manufacturers were faced with the problem of finding an adequate supply of rags to meet their consumption. By the late 1830s, Herbertshire Mill, Stirlingshire, was using twenty-six hundred weight a day. So enormous were the requirements of the industry that the larger mills were closely connected either with regular suppliers or major urban centres where they had collecting arrangements. One Berwickshire mill had agents in Berwick, Wooler, Alnwick, Jedburgh, Kelso and Duns.[27] In 1803, placards in one northern town had this notice pasted on them.

> *To the Ladies*, – Genteel women, who amuse their idle hours in working, frequently throw scraps of linen and cotton of various kinds into the fire. It is requested most humbly, that every lady will reserve these trifles, and direct their maid servants to sell them, because their so doing will prevent £60,000 being annually exported to foreign countries for the importation of old rags to make paper, and which in consequence will become cheaper.[28]

In the 1860s, it was stated that rags were collected regularly from only about half of the houses in the country, and that coloured rags and even waste paper were equally valuable.

> Every housekeeper ought to have three bags; a white one for the white rags, a green one for the coloured, and a black one for the waste paper.[29]

So recycling may be a new term in name only!

In 1830, it was calculated that 24 per cent of paper output was made from imported rags. Thereafter the proportion decreased even though the actual total weight of imported rags increased.

FIVE YEAR PERIOD	AVERAGE PERCENTAGE OF TOTAL PRODUCTION WHICH COULD BE MADE FROM IMPORTED RAGS.
1830–4	20.6
1835–9	18.2
1840–4	11.0
1845–9	10.2
1850–4	9.2
1855–9	9.2[30]

In 1790, 6,200 tons of rags were imported but this figure was not reached again until 1818 due to the disruptions in trade caused by the Napoleonic Wars. Thereafter imports rose slowly to about 16,000 tons around 1860[31] and

20,000 tons in 1863.[32] The export of rags was prohibited from France, Holland, Belgium, Spain and Portugal in order to supply the papermaking industries in those countries. Germany and Italy furnished the principal supplies of linen rags to Britain followed by Russia and Austria. In 1861, the *Paper-Trades' News* mentioned imports additionally from Alexandria, Colombo, the East Indies, China, Australia, the Cape, the Canary Islands and Buenos Aires.[33] Rags were obtained also from the Scandinavian countries, the Channel Isles, South Africa, Egypt and elsewhere. Shippers exporting the industrial products from Britain welcomed something to fill the holds on their return voyages.

ANNUAL AVERAGE in 5 year periods	GERMANY tons	%	ITALY tons	%	RUSSIA tons	%	OTHERS tons	%	TOTAL tons
1800–4	2138	62	542	16	27	1	760	21	3467
1820–4	4066	58	1391	20	846	12	657	10	6960
1830–4	4436	51	2306	26	1104	13	667	21	8715[34]

Hamburg had a rag market of international importance and in the period 1856–60 was averaging nearly 13,000 tons of rags exported to Britain, a share of about 57 per cent compared with Russia's 16 per cent. Rags continued to be imported into

26. Cutting rags by hand, c. 1850.

Britain throughout the nineteenth century and into the twentieth with, for example 21,200 tons in 1882, 23,032 tons in 1892 and 18,692 tons in 1902.

The character and quality of the paper was determined partly by the type of rags chosen, so the selection of rags was crucially important. It was said that their qualities gave a pretty clear indication of the state of comfort and cleanliness of the countries from which they had come. The linen rags of England were generally very clean and required little washing or bleaching before pulping. The Sicilian rags, on

27. Nineteenth century spherical revolving rag boiler.

the contrary, might be so dirty that they would be washed in lime before despatch. The greater portion of the rags from northern Europe were so dark in colour and so coarse in texture that it was difficult to imagine how they could have formed part of any inner garments.[35] Yet it was the linen which was in greatest demand for making high quality paper.

Ways of sorting rags changed little over the years. In London, there were rag dealers, sometimes even the stationers like the Fourdriniers, who purchased rags and employed women and children to sort and grade them. The mills too had their own rag sorters and cutters. Each sorter had a table with the blade of a knife mounted vertically on it. The cloth was cut up by the knife into four inch squares and thrown into one of six boxes depending upon the appropriate grade. This might be determined by the type of cloth, the colour, or the amount of wear and rottenness. Seams and stitching tended to be less worn and therefore would be stronger than other areas of a garment. These parts would require more beating to reduce them to the same consistency as more rotten sections. Unless rags of the same quality were beaten together, the stronger fibres were not pulped to the same extent as the weaker ones so the paper would be uneven. It was better to beat the same type of rags together and blend the different batches later.[36]

Cleaning the rags was mechanised to a certain extent. The rags were fed into 'dusters' in which spikes on a revolving shaft violently tossed the rags so that some dirt was loosened and fell out through gratings. This was a very dirty job through the dust which was raised and was later improved with ideas from the cotton industry in which the dust was extracted by a draught of air. In 1770, there was mention of a rag cutting machine in Scotland but these were really developments of the late nineteenth century and would only cut up into smaller pieces rags which had been sorted, graded and had had the seams removed already. Further cleaning was carried out at an early period in steam heated chest boilers with lime or caustic soda. Later, boiling time might be 3 to 7 hours at pressures of between 15 to 30 p.s.i. in spherical boilers which were rotated to cook the rags evenly. Then would follow bleaching and beating.

Bank notes, bills of exchange and other commercial and legal documents had to be tough to last and therefore had to be made of the best quality. Other paper for writing or printing did not need to be quite so tough and might consist of a mixture of good quality fibre to give strength filled out with poorer quality. Most of the rags used to manufacture writing paper were collected in Britain but those for the best printing paper were imported principally from Italy and Hamburg.[37] Since the raw material might be uneven in its consistency, a great deal of supervision and skill was needed in sorting if a first class product were to be the result. It had long been recognised that the use of rag as the sole basic material for paper was a lasting hindrance to the expansion of the industry if it were to reach its full potential. Papermakers wanted a raw material which would be as homogeneous as possible so they could rely on it to give the grade of paper they required. This problem is well illustrated by the firm of Balston's after the Second World War. Nylon fabrics had been introduced so they had taught their rag sorters how to recognise them. After sizing and being dried on steam cylinders, one batch of paper emerged with what looked like grease spots. Eventually it was discovered that some nylon had been blended with cotton in a fabric which had passed unnoticed by the rag sorters and had melted with the heat from the cylinders. The company was forced to abandon using rags for its high quality Whatman papers.

Within Britain, the continued expansion in the cotton textile industry through the development of the self-acting mule and the power loom helped to alleviate the rag shortage. The price of cotton cloth fell dramatically and people used more of it, so the supply of rags from this source increased. Then there was the waste from spinning and weaving which could be turned into paper which too became an important source. However, during the 1850s, ways were developed of recycling some types of waste cotton on condenser spinning mules and turning it into flannelette sheets or dusters and the like, so one source of raw material for the paper industry became scarcer. Chains replaced ropes for many purposes at about this time too so that source of fibres also became scarcer. The price of rags used in ordinary printing paper rose by 28 per cent from an average of 12 s. 6 d. per cwt in 1848–52 to 16 s. per cwt in 1853–6. Some other qualities of rags doubled in price during these years.[38]

One way in which the provision of rags was increased came about through the introduction of bleaching. In 1774, Karl Wilhelm Scheele of Sweden had discovered what he called 'dephlogisticated muriatic acid', to be re-named 'chlorine' in 1810.[39] In 1785 or 6, the French chemist, C.L. Bertholet, published a work specifically on the bleaching effects of chlorine. James Watt introduced it to Scotland in 1788 and, in the textile industry, its use began to replace the older long and tedious methods. In 1790, William Creech of Edinburgh published an essay for Robert Kerr on the new method of bleaching 'from the French of Mr. Berthollet'.[40] William Simpson at his mill near Edinburgh carried out experiments in 1791, the results of which Joshua Gilpin observed in 1795,

> This improvement was first adopted by William Simpson from Kerr's translation of Berthollet's essay on bleaching. William Simpson after a number of experiments adopts the following mode – the coarse rags are first put in the beating engine for 2 hours, then carried into a larger boiler room where they are boiled in a strong alkali of lime of potash, pearl ash or even in limewater itself – after being thus boiled they are taken out and put in a square box of cast iron with small holes where they are pressed with a screw till dry. When dry they are picked to pieces by women and put into square boxes made very stout and tight of the width of an engine – these boxes have a cover fitting close which after the rags are in must be made very tight by pressing paper at every crevice where the air can escape. A small furnace is built between two of these boxes – on the top of this furnace is fixed a square cast iron boiler in which is placed a leaden retort and a top to the boiler fastened down.[41]

The bleaching chamber was connected with the retort or still, containing black oxide of manganese and common salt, onto which sulphuric acid was poured. The chlorine gas passed off into the chamber and bleached the rags.[42] The chemical reaction must have been difficult to control accurately to produce an even flow of gas. What the working conditions must have been like under these primitive conditions is hard to imagine because the chlorine must have escaped easily from the retorts and chambers and there is evidence that no time was allowed for the gas to disperse before the men had to start moving the rags when treatment finished.

Now in 1792, Clement and George Taylor were granted a patent for using chlorine bleach to whiten rags, and claimed,

> A new method of decomposing or removing all sorts of colours in linnens & cottons, & for whitening all other kinds of linnens & cottons . . . Treat the pulp with a certain amount of pearlash, wash this away, and submit it to the action of dephlogisticated marine acid for a given time in a bleaching machine, afterwards wash the acid away with water.[43]

This description is similar to the process which Simpson was using and so was opposed by the Scottish paper manufacturers. They were supported by James Whatman the younger who had purchased the materials for making trials in 1791 but never actually bleached any rags.[44] Larking and possibly others in the Maidstone area had also been experimenting.

The use of bleaching spread rapidly, particularly after the discovery of chloride of lime and the invention of bleaching powder by Charles Tennant in 1799, because this proved to be easily transportable as well as being much easier to use. He developed its production on a large scale at the St. Rollux chemical works near Glasgow. Output rose and prices fell. In 1805, the works produced 147 tons at £112 per ton; in 1870 over 9,000 tons at £8 10 s. per ton. Bleaching powder proved to be so useful that in 1815 it was claimed that the majority of papermakers employed it.[45]

But bleaching was taken up too enthusiastically, when its disadvantages were quickly apparent, because, while chlorine whitened, it could also degrade and even destroy the fibres. In 1816, James Smith wrote,

> Large quantities of paper, apparently made according to the best process of bleaching, have been rendered unfit for use . . . Different manufacturers are more or less successful in their use of bleaching, but in those cases where the process is so well conducted that the use of the acid can scarcely be suspected, the care bestowed in washing the pulp causes an additional expense almost equal to that of using a more valuable rag without bleaching. The faults commonly attributed to bleaching paper, and which in some degree obviously exist, are that it is not so proper for writing upon, because the size is injured; that in the course of time, it weakens the colour of the writing ink, and its

own whiteness diminishes; that it is apt to break in the folds, and from its little coherence, is easily worn away by friction.[46]

In 1816, of a quantity of Bibles printed by the British and Foreign Bible Society, one was found two years later crumbling to dust, although it had not been used, owing to the nature of the bleaching process. Then an edition of 30,000 books was printed in 1818 of which no perfect copy remained in 1834 because they had fallen to pieces through excessive bleaching with chlorine.[47] Yet chlorine bleach did much to ease the raw material shortage for not only did it remove discolourations in rags but it also whitened some types of dyes and so made coloured rags available for papermaking.[48]

Through the shortages of rags, many people turned to investigating whether paper itself could be recycled. It is probable that waste paper had been used for many years for making boards and in 1813 a machine was patented in England for shredding waste paper preparatory to its remanufacture.[49] However Matthias Koops earlier had tried to make good quality paper from waste paper. The second edition of his book, *Historical Account of the Substances which have been used to Describe Events, and to Convey Ideas from the earliest date to the Invention of Paper* published in 1801, was printed on paper made from re-cycled waste and shows that he had succeeded for its quality is excellent. In this book he wrote,

> I have had the satisfaction to witness the establishment of an extensive Paper-manufactory, since the first of May 1800, at the Neckinger Mill, Bermondsey, where my invention of re-manufacturing Paper is carried on with great success, and where there are already more than 700 reams weekly manufactured, of perfectly clean and white Paper, made without any addition of rags, from old waste, written and printed Paper; by which the publick has already been benefited so far, that the price of Paper has not risen otherwise than by the additional duty there upon, and the encreased price of labour. And it will not be many weeks before double that quantity will be manufactured at the said mill.[50]

In 1800, Koops explained in his patent how he would extract the size and the printing and writing inks from waste paper to make it fit for repulping. First of all, the waste paper had to be sorted because the treatment varied on whether it was printed with English or German ink or whether there was ordinary writing ink on it. Then the size was removed by soaking the sheets in hot water which was followed by boiling in solutions of American potashes and limewater of different strengths to remove the inks. After this, the pulp was thoroughly washed to remove any traces of alkali and finally bleached to give it a brighter colour. Once the size and the inks had been removed, the pulp could be prepared in the usual way in Hollander beaters.[51]

Earlier Koops had published the first edition of his book in 1800 which was printed on paper made from straw. In the August of that year, he applied for a patent for manufacturing paper from straw and other substances for which no specification was enrolled.[52] In the February of the following year, he applied for a further patent in which his processes were described.[53] He proposed to 'manufacture paper from straw, hay, thistles, waste and refuse of hemp and flax, and different kinds of wood and bark, fit for printing and other useful purposes'. The straw, hay or thistles were cut on a chaff cutting machine into two inch lengths, while the wood was reduced to shavings, again in two inch lengths. 'Wood which contains much turpentine or resinous matter cannot usefully and beneficially be made into paper'. Sometimes these materials might be boiled for about three quarters of an hour or fermented for a few days to give them a preliminary softening. The basic process was to soak them in limewater for six or seven days, turning them so the lime was applied evenly. Then followed washing and boiling for a couple of hours in clean river water. Sometimes crystal of soda or potash might be added to improve the colour and texture. After this would be a further washing and boiling and then pressing to remove the water. Sometimes this pressed material might be fermented for several days or it could be made into paper in the usual way immediately, depending upon the type of paper needed.

The paper in Koops's books made from both straw and wood pulp has survived remarkably well over the years, but the straw paper is rather yellow and is the poorest in quality of the three, the waste, the wood and the straw. Koops used the water from the river Thames, because he could most easily procure that where he was experimenting but he supposed that spring water would do as well. Before the new London Bridge was built shortly after 1821, there was virtually no tide in the Thames above that point so Koops would have had fresh water, which presumably in those days was relatively unpolluted. The principles described in Koops's patent are similar to those which are employed today to make paper pulp from wood and similar materials but the whole of his process took a very long time and, with the repeated boilings, would have been expensive in fuel and probably chemicals too.

The reason why Koops tried to make this paper is clearly stated in his first book.

> All Europe has of late years experienced an extraordinary scarcity of this article [rags], but no country has been so much injured by it as England. The greatly advanced price, and the absolute scarcity, equally operate to obstruct many printing-presses in this kingdom; and various works remain, for these reasons, unpublished, which might have provided very serviceable to the community.
>
> The great demands for Paper in this country have rendered it necessary to be supplied from the continent. This supply is extremely precarious, and is likely to be more wanted, as the consumption of Paper increases, because the material, which is the basis of Paper, is not to be obtained in England in sufficient quantity. The evil consequence of not having a due supply of Rags has been the stoppage of a number of Paper-mills; for it is a manufactory which requires numerous hands (of men, women, and children); a great

number of whom have been thrown upon their respective parishes for
want of employment.[54]

In fact Koops's own employees were soon to be thrown out of work in
December 1802 and the pulp was left rotting in the vats because his enterprise
failed disastrously.

Koops not only had the Neckinger Mill in Bermondsey, but in 1801 floated a joint
stock company to build a much larger one at Mill Bank, Westminster, about a mile
from Westminster bridge.[55] Here he established an enterprise that was to be much
bigger than any other papermaking concern of that day. There was one steam engine
of 8 hp installed by John Rennie and another much larger one of 80 hp which cost
£6,000.[56] When the size of a steam engine for driving the average textile mill at that
time was only about 15 to 20 hp, one of 80 would have been immense. The mill had
20 vats making it easily the largest in Britain and probably the world. It is suspected
that the production costs of these new processes proved to be too expensive even
though prices of rags were very high at the time through the blockades imposed
during the Napoleonic Wars. The import of rags was taxed in 1803 but this came
too late to save the enterprise which would have been beset by other problems such
as the high taxes on coal shipped into London. Around £71,525 was subscribed to
establish the mill, of which £45,000 was spent on the 'buildings, machinery and
utensils'. The company went bankrupt with debts of £10,500,[57] and thus ended rapidly what was obviously an over-capitalized, over-ambitious but ingenious project.

Koops's failure did not deter other inventors in subsequent years. One substance
in particular, straw, attracted many people because it was readily available in this
country. In 1824, L. Lambert patented a process similar to that proposed by Koops.
He proposed to treat the straw

> with quick lime or caustic potash, soda or ammonia . . . then use quick
> lime and sulphur to free it from the mucilaginous and silicious matter so
> prejudicial in papermaking . . . then wash it . . . bleach it with chlorine or
> exposure to the open air.[58]

This process does not seem to have been exploited commercially although some
straw was used in subsequent years. In 1854, John Evans of Dickinson's took out
a patent[59] for treating the waste refuse of the Brazilian grass which was used for
platting hats but the supply from the hat factories of St. Albans and Luton was too
small and uncertain to make this a viable source of fibre for papermaking.

It is in John Cowley's patent of 1856 that we can begin to see a process that
might have been able to turn straw into a more popular fibre for papermaking. He
still used alkalis to break down the plant stems and release the fibres. He boiled the
straw under pressure at a temperature of 250° F. inside an iron vessel and continually
circulated the liquor through it. In this way, he could reduce the boiling time to

between 8 to 10 hours.[60] This is the first time these features appear in a patent and it is probable that the earlier patentees did not use them so their pulp took far too long to prepare. Then followed bleaching and beating.

However, while on the Continent straw has become quite a common material for papermaking, in England it has always been regarded with disfavour because its fibres are very short and it tends to make a hard paper with poor tearing strength.[61] There is also the problem of how to dispose of the effluent left over after treatment. In July 1852, John Dickinson wrote to Charles Longman,

> Joynson has been making paper with a large mixture of Straw and says
> it will answer well. It appears that it takes the size well, and handles
> extremely well, and thick.[62]

In the 1850s, Herring commented about straw that it needed a most expensive process for its preparation in which much was wasted. Yet he saw it as the chief alternative to rags. One of the problems in using straw always has been that the nodes at the leaf joints take more time and chemicals to break down than does the rest of the stalk so the pulp may be lumpy. Another problem with straw was the presence of weeds in it.[63]

However straw could be used as an auxiliary to rags and sometimes the mixture might be as high as 50 to 80 per cent. While thick brown paper could be made from it, only an inferior type of paper for printing or writing could be produced. The cost of rags was £17 per ton and straw only £2, but there would be much more waste with straw and, when the cost of the power, labour and chemicals was added, straw pulp cost about the same as rag.[64] Up to 1860, nearly everything which had been proposed as a substitute for rags had been excluded either through its scarcity, or cost of transport, or cost of preparation, or a combination of these. It was a period when many people patented ideas for making paper from a wide range of substances about whom Herring commented,

> Given, plenty of money to work out their processes, sanguine but unpractical
> inventors may, regardless of cost, produce paper from wood, hay, or
> stubble; but, to quote the words of Dr. Forbes Royle, 'The generality of
> modern experimentalists seem to be wholly unacquainted with the labours
> of their predecessors, many of them commencing improvement by repeating
> experiments which had already been made, and announcing results as new
> which had long previously been ascertained.[65]

The situation was about to change dramatically.

ESPARTO AND OTHER
TYPES OF PULP

28. Stack of esparto at the North Wales Paper Company's Oakenholt mill, c. 1905

In the British Isles, the solution to the shortage of rags for making paper was found in a peculiar source. Growing in the hot climate of southern Spain and in North Africa was a grass *Stipa tenacissima* and *Lygeum Spartum*, more commonly called esparto. It had long thin leaves, growing to three or four feet which, to minimise an excessive loss of water by transpiration, curled round into a long tube. The grass was wiry and tough and was harvested by being pulled up by hand. The coarser impurities, such as the roots, were picked off and the rest baled and sent to Britain in the holds of ships which had carried coal for fuelling steam ships in those regions. Without this carriage as a return cargo, esparto probably would not have been an economical source of fibre for papermaking because the true fibres were located in bundles within the leaf and amounted to only about 45 per cent of the dry weight of the raw grass. The fibres were short and fine, in fact the smallest in diameter of the common papermaking fibres. The average dimensions were 1.5 mm. length and 0.013 mm. diameter.[1]

Even though the esparto fibre was so small, it could be beaten to give bulk and opacity to paper and so was used in featherweight printing papers. It produced paper with closeness of texture and smoothness of surface and, when wetted, it expanded less than almost any other fibre. This made it especially suitable for manufacturing high-quality printing papers of all sorts and, after the Second World War, was favoured by Sir Winston Churchill for the publication of some of his books. Also the short fibre length gave clarity to watermarks so it became popular for good quality writing papers too. For added strength, esparto might be blended with a proportion of longer, stronger fibres, such as rag or some woodpulps.

Of course all these advantages were not immediately discovered when esparto was first tried as a pulp. The earliest person to take out a patent for its use was Miles Berry in 1839. He stated that it grew in abundance around the borders of the Mediterranean where it could be collected easily for export.

> To make white paper and card-boards, it is necessary to beat or bruise the material or crush it, then to steep it in lime water of about the strength of two pounds of lime for one hundred pounds of 'Esparto', allowing it to steep until it begins to ferment. It is then put into ordinary pulp engines used in paper mills, and mixed with one quarter, one third, or one half of rags or other fibrous material, according to the strength and quality of paper wanted to be produced; the material being thus reduced, it may be bleached in the usual manner, and afterwards sifted or refined, and made into paper either by machinery or by hand.[2]

For making brown cards, for example for the Jacquard looms, no bleaching was necessary and the steeping took less time.

It does not seem that many manufacturers tried esparto at this time but interest in it must have continued and it was first imported into England in any quantity by Noble in 1851.[3] Two more patents followed, that of Jules Dehau of Paris in 1853 and another in 1854 by James Murdoch.[4] A number of makers, including the Potters of Darwen in 1856, experimented in vain with it, but Thomas Routledge took out the key patents in 1856 and 1860.[5] He recommended in his 1856 patent treating the esparto

> with a caustic ley, composed in the ordinary manner of soda or potash and lime, but containing an excess of lime, that is to say more lime being present than is necessary to render the alkali caustic, this being necessary to bring the gums resinous and siliceous matter (coating and cementing together all raw vegetable fibres more or less) to a soluble state.

It was probably through using a strong solution that Routledge achieved success. Boiling in an alkaline solvent of lime lasted for 3 to 5 hours followed by washing in carbonate or bicarbonate of soda, bleaching and beating.

Probably Routledge was fortunate in the timing of his experiments for the price of soda fell during the first half of the nineteenth century. Leblanc invented one process for manufacturing soda ash in 1791, and the price rose from about £45 per ton in 1800 to over £60 in 1810 and then fell to £35 in 1820. By this time, an alkali industry had grown up and prices continued to drop to £10 in 1840 and £7 in 1860. Routledge operated Eynsham Mill, Oxfordshire, where he experimented with esparto and sent samples to the Society of Arts.[6] From about 1858, Eynsham Mills were producing paper from esparto alone. Experiments were conducted at the Bridge Hall Mill, Bury, Lancashire, of James Wrigley and Son in 1860 and again in 1861. Routledge was not present when the first trials failed to reach production costs of £24 per ton. Wrigley invited Routledge to superintend the second set of experiments and admitted that a very satisfactory pulp was made although the amount of alkali and chlorine needed in the process seemed to him to be so great as to make the use of esparto excessively costly for it needed ten times as much chlorine gas in the bleaching.[7]

In the meantime, John Evans and the Dickinsons also had been experimenting with esparto. However, the amount of soda employed and the impure effluent obtained made its preparation in their mills to the north of London on the river Gade impossible for they were legally bound not to contaminate that river or the Colne any more than they did already. Therefore they linked up with Routledge in 1860 and established a subsidiary company under his practical management for the production of esparto half-stuff at the Ford Works at Ford, near South Hylton in Sunderland. The initial accounts of the subsidiary firm, issued on 16 October 1865, showed that the first year was spent in experiment. In nine months, 50 tons of paper were produced of which the bulk was too coarse for anything but newsprint. In the second year, it was hoped to produce 120 tons of half-stuff in every six-day week. The half-stuff was sent to the Dickinson mills for making into paper in hydraulically pressed round, upright canvas-wrapped bales of about 3 cwt. each for bleaching and processing into whole-stuff. Such was the Dickinson's consumption of esparto that in 1877 they leased both Fourdrinier's old mill at Frogmore as well as the Two Waters Mill where esparto half-stuff could be prepared.[8]

At the International Exhibition in 1862, the jury awarded Routledge a medal for his paper made wholly from esparto and this caused other papermakers to become interested in its possibilities. John Evans, in a paper read at Edinburgh in 1863, saw either straw or esparto as the only alternatives to rags but was afraid that supplies of esparto would be limited and, if cultivated, would become too dear.[9] While the price did rise quickly at first, additional supplies were found in Tunisia and esparto became an important raw material for the paper industry in Britain. It had an effect on other industries too, particularly the chemical industry. In 1863, the import of around 19,000 tons of esparto increased the consumption of soda ash and bleaching powders by around 4,000 tons per annum.[10] Mills with easy access to harbours, such as those in North East England, changed to esparto. This was also an important area of the chemical industry.

The Dee estuary near Chester was another location favoured with good shipping facilities, supplies of coal and also chemicals. In 1854, the London firm of stationers, Grosvenor Chater, had purchased the Abbey Mill at Greenfield, North Wales, because there was a never failing supply of good water from the spring at St. Winifred's Well, coal for steam raising from the near-by Englefield mine and also chemicals for bleaching from the firm of Muspratt Bros. and Huntley at Flint as well as transport by sea and rail.[11] In 1865 this mill quickly switched to esparto which was found to be very suitable for making the stationery papers in which the firm specialised. Close by the North Wales Paper Company built a mill in 1870 on a virgin site at Oakenholt, expressly for making paper from esparto. This was financed by M'Corquodale, the Liverpool firm engaged in printing railway timetables and the like. One of the reasons for the choice of this location was the proximity of the alkali works of Smith and Mawdsley as well as the near-by collieries. Oakenholt specialised in the production of newsprint and white and coloured printing paper.[12]

The use of esparto spread quickly in Scotland where its qualities were appreciated for the paper which these mills supplied to the printing industry. In 1861, nine mills in the Esk area used some 148½ tons of esparto with 6,565 tons of rags and by 1863 the esparto figure had risen to 2,215 tons.[13] In 1870, the forty Scottish mills enumerated in the River Pollution Report were using over 44,000 tons of esparto, compared with just over 18,000 tons of rags and less than 2,000 tons of other materials. With few exceptions, only those mills producing cartridge and wrapping paper still relied exclusively on traditional raw materials and seven mills had changed entirely to esparto.[14] Even early in the present century, Scottish papermakers were still relying considerably on esparto for in 1908, that country took 129,910 tons out of a total imported of 192,975 tons, or approximately two-thirds.[15] The quick increase in consumption of esparto and its importance to the industry may be judged by the quantity imported into Britain shown by the records at the Custom House.

YEAR	QUANTITY IN TONS
1861	16
62	876
63	19,326
64	43,403
1865	52,324
66	70,041
67	55,074
68	95,880
69	87,442
1870	104,870
75	141,900
1880	191,229
85	201,036
1890	217,028
95	186,408
1900	200,280
05	191,114
1910	193,218
1914	183,114.[16]

The esparto grass reached the mills in huge bales containing bundles of grass tightly packed together by hydraulic pressure. The bales were broken open and the bundles loosened by women who sorted and began to clean them. Then a machine called a 'willow' was employed to beat out all the dust and dirt and to discharge the esparto onto a travelling belt where the roots, weeds, etc. might be picked out. This material was fed into digesters, holding at first two or three tons but later about four, which were based on the ideas contained in John Cowley's patent. In fact his, or Sinclair's 'vomiting boiler', proved to be more suitable for esparto than for straw because the esparto leaves did not bunch up as much as the stalks of straw so the liquor circulated more evenly. These boilers were large drums, set vertically, latterly about 9 to 10 feet diameter and 14 feet high (2.7 x 3 x 4.2 m.). The esparto was fed in through a cover plate in the top, the top secured and the steam turned on. With a pressure of about 20 p.s.i. in the early days, later raised to 40 or even 80 p.s.i.,[17] boiling occupied four to five hours. This was where the patent of Routledge was important because he recommended adding to the ordinary soda or potash more lime than was necessary to make the alkali caustic. This rendered the non-cellulose material, mainly pectic carbohydrate, soluble and dissolved all the resinous and silicious material which bound the cellulose fibres together.[18]

The caustic liquor, which had turned black, was drained out of the digester into storage tanks where it was treated to recover as much of the chemicals as possible. Routledge had begun to investigate ways of 'utilization of certain products after boiling esparto' in patents taken out in 1865 and 1866. By boiling down the liquors and destroying the organic matter by heating and burning, he treated the remainder with sulphate of soda and calcination or with carbonate of lime in order to recover the caustic soda.[19] However recovery processes do not seem to have been very effective until the 'Porion' evaporator was devised in 1877 and in 1886, the 'multiple-effect' apparatus was invented. In this, steam would boil off the liquor as the pressure was reduced and this steam could be used to heat a further batch of liquor which had been cooled previously, hence the multiple-effect. About 75 to 85 per cent of the soda could be retained and used again but it was the disposal of this remainder and of the organic matter which caused great pollution in rivers and estuaries and was to lead to the abandonment of esparto in the 1950s when pollution laws became stricter. Sometimes the esparto was washed in the digester and this water also might be treated to recover the soda.

Then the esparto was taken out of the digester through a side cover which was a most unpleasant job with the hot steam and fumes from the caustic soda. Washing followed in a form of Hollander beater called a 'potcher' to remove the rest of the organic matter. Here the grass was broken up so that all the soda could be washed out. At this point, Routledge recommended rinsing the fibre in a solution of carbonate or bicarbonate of soda. The pulp might be bleached in the potcher or this could be done in tall bleaching vessels in which the bleaching liquid was circulated.[20] Washing to remove all traces of bleach was imperative. Often at this stage, the esparto was formed into 'presse-pâte', a sort of thick board, on a machine rather like the wet end of a Fourdrinier paper machine, so it could be further cleaned, dried and sent out as 'half-stuff'. Then the esparto pulp could be treated like any other half-stuff and beaten ready for making into paper.

Towards the end of the nineteenth century, the sharp rise in imports of esparto slackened. While this was largely due to the growing challenge of wood pulp, esparto had probably found its natural level because it was limited in its application through the shortness of its fibres and through the costs of its preparation. During the First World War, there was difficulty in obtaining esparto and some mills had to switch to straw instead. During the Second World War, virtually no esparto was imported and so once again recourse was made to straw. The consumption of straw for papermaking in the United Kingdom rose from 127,000 tons in 1941 to 346,000 tons in 1945. This declined to about 100,000 tons in 1950 when the use of esparto had increased to its pre-war proportions at 307,000 tons.[21] Thereafter the consumption of esparto began to decline and the esparto pulp preparation plants closed down as they could not operate economically, partly through the need for costly effluent treatment. Some esparto pulp began to be imported from those countries where the grass grew but even this became uncompetitive as the grass was always harvested by hand. Today esparto, once a peculiarly British fibre for papermaking because no other country in Europe or North America used it, has almost disappeared from the papermaking scene.

Imports of Esparato

YEAR	ESPARTO GRASS Tons '000	ESPARTO PULP Tons '000
1938	316.2 (about)	–
1954	315.4	–
1955	308.3	–
56	270.4	–
57	191.5	–
58	187.0	–
59	193.2	–
1960	233.5	9.6
61	214.7	8.1
62	197.3	4.9
63	179.6	3.2
64	149.1	7.5
1965	129.0	8.9
66	98.6	15.4
67	65.6	20.7
68	50.2	23.3
69	40.0	20.5
1970	20.7	19.1
71	18.2	15.7
72	4.2	16.2
73	3.5	14.6.[22]

Wood as a Fibre Source

As far back as 1719, Réaumur had realised the possibility of making paper from wood through watching wasps. From that time on, many people tried to follow the example of these insects, including Matthias Koops, but no one succeeded in devising an economical method. Around 1826, a Mr. Sharp, who had a mill in Hampshire, took out a patent for manufacturing paper from pine shavings.[23] Then in 1838, there was a further patent by Desgrand for making paper and pasteboard with wood reduced to a state of pulp. He proposed taking logs 4 to 6 feet long and splitting and cutting them up into small chips, removing the knots and soaking them in water saturated with lime for three to six weeks in a warm temperature. This would soften the glutinous parts of the wood, which could be washed away, leaving the fibres.[24] While this may be seen to be a forerunner of later chemical processes, this method would have been far too slow to produce pulp in the quantities needed by the industry.

In modern papermaking, it is not the bast fibres as in the case of the mulberry which are used but the main part of the tree, the heart and the sapwood. Usually the discoloured bark is removed for steam raising unless a brown paper is being made. The tree grows by the *cambium* cells immediately beneath the bark dividing and forming new cellulose fibres. In course of time the new cells become *lignified* by a complex process of adding *lignin* around the original cellulose. This new growth occurs quickly in the spring and slows down during the rest of the growing season, thus forming the rings which can be seen in the cross-section of a log. The fibres contained in these rings may differ according to the growing periods and also may change with the age of the tree so that the heartwood becomes darker through the formation of tannin and gum which render it less permeable to liquids.

For papermaking, trees are divided into two main classes, softwoods and hardwoods. In the former class are cone-bearing trees such as the pines, spruces, firs, hemlocks and larches. About 95 per cent of the volume of a typical softwood

1000 POUNDS
REWARD.

The Proprietors of a leading Metropolitan Journal OFFER the above REWARD to any person who shall first succeed in

INVENTING OR DISCOVERING

the means of using a

CHEAP SUBSTITUTE

FOR THE

COTTON & LINEN MATERIALS

NOW USED BY

PAPER-MAKERS,

Subject to the following conditions:

1. The material must be practically unlimited in quantity, and be capable of being converted into pulp of a quality equal to that which is at present used in manufacturing the best description of newspaper, and at a cost, *cæteris paribus*, not less than ten per cent. lower.

2. It must be tested, approved, and adopted by three eminent manufacturers of paper (two of them to be named by the advertiser), whose certificate shall entitle the inventor to the payment of the reward.

3. This offer will be in force only for a period of 12 months from the 20th of May, 1854.

Apply by *Letter* to A. B., Messrs. SMITH & SONS, 136, STRAND.

29. *The Times* advertisement for a reward for new papermaking materials, 1854.

consists of fibres with which paper can be made. In the Norway spruce, for example, the fibres are on average 3–3.5 mm. long. There are also occasional resin canals and wood rays which run from the outer layer of the cambium towards the centre of these trees and form part of the complex transport and storage system for plant nutrients. However, they contain much of the wood resin that later becomes pitch and causes trouble in the papermaking processes. Hardwoods are more complex and varied in structure. The sap is carried by a series of vertical vessels with comparatively wide, thin-walled, open-ended cells, connected together with the wood rays. The proportion of fibres depends considerably on the species of tree but usually averages 55 to 70 per cent by volume. The fibres are short, only 1–2 mm. and vary in thickness according to the species. Very dense woods are rarely used for papermaking because their fibre walls are thicker.

To make paper, the cellulose fibres must be freed from the surrounding constituents because it is only the cellulose which has the property of forming the bonds so essential in papermaking. Cellulose is mostly resistant to the action of

30. Late nineteenth century machine for grinding wood to make mechanical pulp.

chlorine and dilute sodium hydroxide under mild conditions. Lignin on the other hand has no fibre forming characteristics and is attacked by chlorine, hypochlorites, sulphur dioxide and sodium hydroxide when it forms soluble derivatives. Therefore the cellulose fibres can be released from the wood either by a straight disintegration to form mechanical wood pulp, in which case the lignin still remains in the pulp, or the lignin can be dissolved from around the cellulose by chemical means and then washed away, making chemical wood pulp. Sometimes today there is a combination of both processes.

Mechanical Wood Pulp

In 1840, Friedrich Gottlob Keller, a German weaver in Hainichen, Saxony, secured a German patent for grinding wood to make paper. He broke up blocks of wood into its fibres by pressing them against a revolving wet grindstone. The first ground-wood paper was made by K.F.G. Kuehn at Alt-Chemnitz as newsprint with a mixture of forty per cent rag fibre to give it strength. In 1846, Heinrich Voelter, a papermill director in Bautzen, Saxony, bought the Keller patent and devised practical machines for quantity production. These machines were built by I.M. Voith in Heidenheim, Württemburg, whose company became important manufacturers of machinery for the paper industry. For five years Voelter devoted his time entirely to the practical and commercial promotion of the new process.

By 1852, ground-wood pulp was being used regularly in some German mills. The first of this pulp produced commercially on the American continent was in 1867 when Albrecht Pagenstecher founded a pulp mill at Curtisville in Massachusetts. Output here was half a ton a day. Newspaper publishers were reluctant to adopt the new material because it was considered shoddy and an inferior stock. Eventually they discovered that such paper had good printing qualities and that it lowered the cost of their white printing paper. In the United States of America, the price of newsprint in the early 1860s was about 25 cents per pound. Ground-wood pulp was first sold at 8 cents per pound and soon dropped to 4 or 5 cents. During the 1890s, it fell to only one cent and reduced the price of newsprint from 14 cents per pound in 1869 to two or less in 1897. In the matter of cost alone, ground-wood pulp effected a revolution.[25] The first was imported into England around 1870.

To make ground-wood pulp, the timber was first prepared by generally being de-barked for the bark required too much treatment to whiten it and caused too many impurities. The logs would then be cut to the required size to fit into the grinding machine. Essentially the process was to force the logs against the periphery of a rotating grindstone and wash the chips or fibres away with water. By the turn of the century, the grindstone was made from sandstone, about 54 in. (1,375 mm.) diameter and 27 in. (685 mm.) thick. Some were set vertically and others horizontally and by 1900 some grinders were driven by water turbines or other motors developing over one thousand horse power.[26] The grindstone revolved at about 120 r.p.m. within a heavy iron casing provided with several chambers or pockets into which the logs were thrown. The sides of the log came into contact with the stone against which they were forced by hydraulic or other pressure. Cross-grain grinding at the ends of the logs mutilated and shortened the fibres and gave a pulp with different characteristics. In more modern machines, the logs are dropped into a chute and pulled down against the grinding stone by continuous chains and the sandstone has been replaced by stones with carborundum or other compositions. In either case, the stones had methods of re-cutting their faces to keep them sharp.

The chips of wood or bundles of fibres pulled off the logs were washed away in a stream of water. Probably at first sufficient water was poured in to maintain a normal temperature because the grinding process generated heat. This produced 'cold-ground' pulp and the pressure forcing the wood against the stone was kept low. Later the shower of water was reduced deliberately and the wood forced harder against the grindstone to make what was called 'hot-ground' pulp in which a very high temperature was developed at the surface of contact. Hot grinding proved to be more economical than cold grinding and was stated to yield a finer pulp with longer fibres. On the other hand, cold-ground pulp possessed a greater 'wetness', so it would be slower to drain on the paper machine but would have stronger bonding qualities. The heat produced by the friction began to soften the lignin and it was this which enabled the fibres to be separated more easily, leaving them longer. In fact, lignin softens at around 160° C. and we will see later how this has been used to advantage in improving the quality of mechanical pulp within the last few years.

The pulp coming off the grinders might contain splinters or other fairly large chips of wood. These had to be screened out and reprocessed. The pulp itself might be further screened to grade it according to fibre size. The coarser grades might be refined again at the pulp mill. To send the pulp away to other paper mills, it was generally made into boards on a form of cylinder mould machine. However breaking down these boards at the paper mills caused a further deterioration in the pulp and there was the additional expense of drying them and then soaking them again. Yet to transport the pulp in a wet condition would have incurred higher transport charges with the attendant risk of attack by moulds or micro-organisms. The tendency today is for this type of pulp to be used in integrated mills where the wood is turned directly into paper.

The qualities of mechanical pulp could not be modified subsequently in the beating process as could other types so the paper-maker was entirely dependent upon the conditions adopted in grinding.[27] It could not be properly fibrillated or wetted and in consequence the cohesion of the paper made from it was poor. In mechanical pulp, the fibres were inflexible, short and possessed very little of the bonding properties that could be obtained in other pulps and so for newsprint often at least 15 per cent of different stronger fibres had to be added. Also this pulp was inferior in its chemical constitution because it retained all the resinous and gummy material as well as all the lignin in the original wood. Therefore papers containing it could not be relied upon to keep their colour, strength or texture for any long period. So the presence of mechanical wood pulp in a paper favours the conditions for gradual decay and ultimate disintegration and the period for complete destruction is inversely proportional to the percentage of ground-wood in the paper.[28]

The normal type of tree from which ground-wood pulp was made was the spruce. The colour of its fibres was already reasonably white and so required little bleaching. Since the pulp retained the whole of the lignin, bleaching with chlorine or allied reagents was used only in exceptional circumstances. Some brightening with peroxide is now well established but this is done normally only in integrated mills. It will be evident that mechanical ground-wood pulp is limited in its usefulness but because it was cheap to produce, it became very popular for newsprint and the less expensive grades of wrappings and printings which were not expected to have long lives. The demand for newsprint in particular created a large demand for mechanical wood pulp which gave rise to the great pulp industries on the North American continent as well as in Scandinavia.

Thermo-Mechanical Pulp

By the beginning of the twentieth century, a distinct variety of mechanical pulp was being prepared through grinding logs which had been thoroughly softened by steaming under pressure. This was known as 'brown-mechanical' or 'leather-board' pulp. The steaming attacked the lignin of the wood and began to break it down so that when it was being ground, there was less disintegration of the fibres. This gave a pulp with longer fibres but the result of oxidation of some of the constituents by the steam produced a brown colour which could not be effectively removed. So while this method resulted in a stronger pulp, its colour restricted its use to the manufacture of 'leather-boards' and brown wrapping papers. A lighter coloured pulp was produced by boiling the wood and then steaming it under pressure.[29]

A much more recent process is to make 'refiner mechanical pulp' in which the wood is first cut up into chips. These are fed into refiners which are electrically driven machines in which a pair of discs with blades on their surfaces rotate. The chips are immersed in water and fed into the centre of the discs. The rotation of the discs at high speed breaks up the chips into match-like pieces which are, presumably, heated and softened by the repeated, rapid compressions they receive from the disc blades. Two or three stages of refining may be necessary so that by the time the wood reaches the periphery of the final set of discs, it is broken down into its constituent fibres. The advantage of this process over the usual grinding is that it is not necessary to use logs and that the pulp appears to contain a higher fibre length. However energy consumption is higher.[30]

148

In the late 1960s, 'thermo-mechanical pulp' was introduced in Canada. In this case, the wood is cut up again into small chips which are passed through usually two stages of refining. In the first, they are fed under pressure into water which is heated by steam to a high temperature, around 110 to 140° C. Here they begin to be broken up as the lignin is melted. This first refining process absorbs a great deal of energy which generates more steam. This is bled off and recirculated to heat the incoming chips or is taken away to heat the steam drying cylinders on the paper machine. Therefore this method is proving to be very economical as the heat is re-cycled but of course can be employed only in an integrated mill to achieve its maximum efficiency. The half-stuff is then passed to the second set of refiners where it is broken down into the final pulp. At this stage, the process is not necessarily carried out under pressure. After screening where any large lumps or splinters are taken out for reprocessing, the pulp is washed. Not only does this system increase the strength of the paper made from the pulp but it could lead to an increase of up to 25 per cent in the quantity of paper that can be obtained from a given tonnage of wood. It also, by upgrading the quality of the pulp that can be made from the 'trimmings', such as the branches, bark and roots, increases the yield per tree, a particularly important consideration in the Scandinavian countries where forest resources are becoming steadily more and more limited.[31] The first thermo-mechanical pulp refiners in Britain were installed at St. Anne's Board Mill, Bristol, in 1979.

Chemical Wood Pulp

It has already been shown that the cellulose fibres in wood can be separated by dissolving chemically the lignin and other material that bind them together. Pulps containing most of the wood lignin consist of stiff fibres that do not produce strong paper which deteriorates in colour and strength quite rapidly. Improvement of these properties is achieved by the removal of most or all of the lignin with solutions of chemicals. While the ground-wood process gave the paper industry a cheap pulp in vast quantity, there remained the necessity for a more durable and lasting paper which would have the strength lacking in esparto pulps. Although various people such as Koops in 1800 and Desgrand in 1838 tried to make paper from wood by separating the fibres chemically, it was not until the 1850s that economical processes began to be developed.

The Soda Process

The earliest process that showed promise of success was originated by Hugh Burgess and Charles Watt at Boxmoor Mill, Hertfordshire, in 1851. Part of a weekly edition of the *London Journal* was printed on their paper and proved its practicability.

However interest in England was insufficient to support its development and the inventors left England for the United States where they secured a patent in 1854.[32] Their patent consisted in boiling chips of wood in caustic alkali at a high temperature which became known as the soda process. Boiling would last for 8–10 hours at a steam pressure of around 100 p.s.i.[33] Pulp mills were established first in Maylandsville and by 1863 at Manayunk near Philadelphia with a productive capacity of twenty tons a day which was considered a large volume at that time. Owing to the powerful action of the caustic soda, virtually any type of wood, including the knots or rotten wood, could be reduced to pulp, so that this process could be applied to resinous pines and hard woods but more generally poplar and aspen. Although the fibre was long, strong and bulky, the pulp had a brownish shade and was difficult to bleach. Compared with esparto and straw, the amount of caustic soda needed to be greater and the pressure for boiling higher[34] and so this method probably was more expensive.

The Sulphate Process

The sulphate process was a variation of the soda process. In the process of soda recovery, sodium sulphate, which was cheaper than soda, was added to make up the loss of soda and the liquor comprised a mixture of caustic soda, sodium sulphide and sodium sulphate.[35] This method was invented by Carl F. Dahl in 1884.[36] By it almost any wood species could be pulped, including the resinous pines, and it yielded a pulp which was stronger than when the same wood was reduced by an acid process. Sodium sulphide and sodium hydroxide were relatively expensive but the process was successful through the possibility of concentrating the waste liquor and burning it as fuel in boiler furnaces. The furnace ash contained sodium carbonate and sulphide which could be easily reconverted into the active caustic soda and sodium sulphide used in the cooking process. While there was a net loss of chemicals, this was balanced by the saving in fuel.

First of all, the wood was chipped and stored in hoppers because most chemical pulps have been made in batches until recently. The chips must be fed into digesters where they would be immersed in the liquor and then be heated by steam under pressure. Around 1900, horizontal rotary cylindrical digesters, 22 feet long and 7 feet diameter (6.7 x 2.1 m.), seem to have been the most popular type in which the chips were steamed for thirty hours.[37] Then followed vertical digesters, up to 80 feet high (24.3 m.), in which the steam might either be injected directly, when the liquor gradually became weaker with the increase in water, or there might be a heat exchanger as the liquor was circulated. Cooking conditions, liquor concentration, pressure and length of time, would all be precisely controlled according to the type of wood and the pulp needed. At least once, the digester had to be 'relieved' during the heating up period, when a relief valve was opened and any air expelled and the volatile vapours from the wood, such as turpentine, might be collected and recovered. More recently various types of continuous digesters have been designed in which the chips and liquor are fed in at one end and pass out cooked at the other.

When the cook has been completed, the contents of the digester are blown out by the pressure into a 'blow tank' which helps to separate the fibres in the very soft chips. The liquor, now black, contains most of the lignin and other substances in the wood. The steam passing off during the blowing off is used to heat the liquor and concentrate it so it can be burnt in the furnaces. The pulp is washed, and even this liquor is collected and its chemicals recycled wherever possible. Today, about 85 per cent of the chemicals are recovered but some are lost in washing water that is too weak to justify recovery. The pulp is screened, cleaned and passed either to be made into thick sheets ready for despatch to other mills or, in an integrated mill, stored ready for the paper machine.

The Sulphite Process

As the alkali process for treating wood and esparto marks the period 1850–75, 'bisulphite' pulp is the main one between 1875–80, with both adding to the papermaker's range and variety of materials available. The story of the sulphite process began in Paris in 1857 when Benjamin C. Tilghman was making some experiments with fats in sulphurous acid. The solution was kept in wooden barrels, and to indicate the depth of liquid, small holes were bored a few inches apart. The holes were closed by removable conical plugs made of soft wood, the ends of which, after a period of immersion, became soft and fuzzy. At the time, Tilghman paid little attention to this chemical reaction, but a few years later, on visiting a pulp mill at Manayunk, he recalled the experience. Tilghman then began experimenting and found that a solution of sulphurous acid, kept at a high temperature and pressure, dissolved the material binding together the wood fibres, but the fibres became red and had to be bleached. Analysis showed that some of the sulphurous acid had been converted into sulphuric acid and that this caused the trouble. By adding sulphate of lime, production of the sulphuric acid would be prevented.[38]

After two years work and considerable expense, Tilghman was forced to give up because the acids dissolved his iron digesters which he could not line adequately so they leaked. These ideas were taken up by Carl D. Ekman who was manager of the Bergvik pulp mill in Sweden which was owned by a London company. In 1872 he successfully produced pulp made with a solution prepared from bisulphite and magnesia. He moved to England where he was joined by George Fry who also had been experimenting with the sulphite process. Together they introduced the

Ekman-Fry process. Sulphite pulp was used in England on a fairly large scale by 1880 and mills were established at Ilford in Essex and Northfleet in Kent, working under their patents of 1881 and 1882.[39] In Britain, the first printing paper made from sulphite chemical pulp came off the Northfleet machine in 1886, but in 1884, Ekman had moved to America to the Richmond Paper Company near Providence where sulphite pulp was soon being prepared.

Although with both alkali and acid processes there is degradation of the cellulose fibres, this tends to be more severe with an acid, yielding a shorter fibre length. Therefore, to dissolve the lignin, the wood had to be easily and quickly penetrated by the acid and so the bark had to be removed and also the knots. At first these were sometimes bored out. Then, after chipping, the wood might be further disintegrated by means of crushing rolls. Any knots and unsound portions were sorted out before transfer to the digesters. The digesters themselves were similar in construction to those used in the sulphate process but they had to be lined to prevent the acid attacking the iron. To begin with, lead was used but, with the difference in rates of expansion as the two metals were heated up, the lining tended to 'creep' and so to crack. Various ways of overcoming this were tried but the lead linings were superseded by non-metallic protective coatings composed of 'cements' of various compositions. The introduction of these linings dated from 1883 when Brungger observed that an iron pipe temporarily used for the steam supply to a digester was rapidly coated with a protective scale derived from the bisulphite liquor. This type of coating gave way to a mixture of Portland cement and silica of soda, or ground slate and silicate of soda or a mixture of ground slate and glass with Portland cement. Nowadays, a lining of special bricks or tiles is used.[40]

Two systems for cooking the chips evolved, the Mitscherlich or 'slow-cooking' and the high pressure Ritter-Kellner or 'quick-cook' method. In the 'quick-cook' process, tall digesters, which around 1900 might be 50 feet (15.2 m.) high and lined with acid resisting brick, were filled with wood chips, the liquor run in as quickly as possible and the top clamped down. High-pressure steam was injected straight into the digester and the cooking continued for eight to ten hours or sometimes a little longer. The pulp was blown out of the digester in a way similar to that described for sulphate pulp and treated in large washing tanks. There was no recovery of the waste liquors in this process and in 1910 it was calculated that not only was one-half of the weight of wood put into the digester lost but that also about 300 lbs. of sulphur were thrown away for every ton of pulp produced.[41] However, in this method, the pulp was more thoroughly cleared of the lignin than in the slow-cook process and was freer but was not so strong.

In the slow-cooking system, the pressure was seldom above 45 p.s.i. and the boiling continued for 36 to 48 hours. Heating was carried out indirectly through heat exchangers so the acid was not weakened. Normally the pulp had to be shovelled or raked out of the digester after the pressure had been reduced.

Once again washing followed. The yield from the Mitscherlich process was slightly higher than from the other. In both cases, the pulp could be easily bleached but careful washing was essential to ensure that all acid was removed otherwise deterioration would set in later.

In the sulphite process, the problem has always been what to do with the waste liquor. At first it was discharged to waste but this caused pollution in rivers and lakes. It is difficult for a number of reasons to recover the chemicals and heat from the calcium bisulphite solution, as can be done in the sulphate process, because the liquor cannot be evaporated and burned without the deposition of calcium compounds. Some of the chemicals can be recovered, such as lignin, vanillin and alcohol, but it is the disposal of the rest which has caused this system to fall into decline at the present moment.[42] With any process, there will always remain the problem of what to do with the residues which will cause some form of pollution.

In order to retain the advantages of the sulphite process with the light colour of the pulp and its ability to swell rapidly on beating and refining, different chemicals have been tried to avoid pollution and save chemicals as costs of both fuel and chemicals rise and pollution laws become stricter. Cooking liquors have been developed containing magnesium bisulphite or ammonium bisulphite. Their acidity is much lower than that of the acid sulphite liquor and consequently the retention of hemicellulose and yields are higher and a wider range of woods can be used. In the case of the magnesium base process, the black liquor can be treated in a similar way to the sulphate process. Magnesium oxide is produced in the furnace ash and sulphur dioxide in the gases, both of which can be recovered. The valuable constituents in the black liquor from the ammonia base process form nitrogen and sulphur dioxide in the furnace, of which the sulphur dioxide is used again.[43]

Today there is a wide variety of processes combining features of both mechanical and chemical wood pulping methods. Two stage processes are used in some mills. In one entirely chemical process, the Stora, the chips are cooked first in a mild liquor containing sodium bisulphite and sodium sulphite which is slightly acidic. Then, by adding sulphur dioxide, the liquor becomes strongly acidic for the second stage. In another method, a slightly alkaline cook is followed by an acidic cook. In 'semi-chemical' pulps which yield between 60 to 85 per cent, the chips receive a mild chemical treatment before being defibred in refiners. Then there are 'chemi-mechanical' pulps with a yield of 85 to 95 per cent in which the chips are treated rapidly with nearly neutral sodium sulphite liquor before refining in a modification of the mechanical refiner process.

Throughout the end of the nineteenth century and for much of the twentieth, almost all the wood pulp was imported. The first year that such imports were separately classified was in 1887.

British Imports of Wood Pulp

YEAR	CHEMICAL	MECHANICAL	TOTAL
	Tons	*Tons*	*Tons*
1887	N.A.	N.A.	79,543
1890	N.A.	N.A.	137,837
1895	N.A.	N.A.	297,095
1897	179,525	225,317	388,304
1900	N.A.	N.A.	487,742
1905	249,468	328,544	578,012
1910	371,715	489,158	859,873
1913	411,803	565,954	977,757[44]

From 1938, the import figures have been

Imports of Wood Pulp 1938–1973 in 000 tons

YEAR	CHEMICAL	SEMI-CHEMICAL	MECHANICAL	TOTAL
1938	970.6	–	673.2	1,643.8
1954	1,093.3	–	614.1	1,707.4
1955	1,325.8	–	689.6	2,015.4
1960	1,701.9	–	757.3	2,459.2
1963	1,637.0	19.1	696.3	2,352.4
1965	1,744.5	35.3	840.4	2,620.2
1970	2,103.0	19.3	616.1	2,738.4
1973	1,685.4	19.6	432.4	2,137.4
1976	1,601.4	15.2	267.7	1,884.3
1980	1,381.3	12.0	183.6	1,576.9
1981	1,381.7	10.7	121.7	1,513.5
1982	1,160.2	5.2	94.7	1,260.1
1983	1,277.2	7.9	115.0	1,400.1
1984	1,307.4	10.7	226.3	1,544.4
1985	1,256.6	8.4	182.8	1,447.8
1986	1,389.3	6.0	194.4	1,589.7[45]

These figures show the enormous growth of the paper industry which has been achieved through developing the various types of wood pulp. It is certain that not enough rags could have been found to meet this demand even if it had been feasible to process such a large quantity, because rag sorting relied upon people to make the selection. However, wood pulp had its problems. The mechanical retained the lignin and other substances which made it deteriorate quickly and also it lacked strength. The sulphate pulps were stronger but were difficult to bleach. The sulphite pulps had to be washed thoroughly to ensure that all acid was removed and also the effluent presented disposal problems to prevent pollution. The struggle still continues to find an economical way of making a pulp that will be cheap, have a long life when it is made into paper and in no way harm the environment.

THE PAPER MACHINE MATURES

Growth During the Nineteenth Century

The paper machines changed the face of the industry during the nineteenth century. Output rose dramatically as more machines were introduced and their performance was improved, but the numbers of mills declined and their locations changed. To begin with, the numbers of mills actually increased in England and Wales from around 430 in 1800 to 564 in 1821. Scotland had under 50 mills in 1800 and 74 in 1823. However the decline soon set in for in 1831 there were only 507 in England and Wales. In 1850, there were 327 mills in England and Wales and 51 in Scotland and these numbers had deceased further to a total for both countries of 350 in 1873, with a greater decline in England than north of the border. In 1890, there was a peak of 69 mills in Scotland but by 1900, the numbers had shrunk to 221 in England and about 50 in Scotland.[1] It was the smaller mills where paper was made by hand that had disappeared because they were unable to compete with machines.

In 1810, about six machines were at work compared with 760 vats.[2] By 1822, Bryan Donkin had built 42 machines, to which must be added those of John Dickinson. In 1830, there were only some 480 vats but still under one hundred machines. In 1837, figures for machines vary widely from Fourdrinier's of 279 to

31. The papermachine in 1850.

32. The wet end of a late nineteenth century papermaking machine at Oakenholt, c. 1905.

Spicer's at only 105, with the probable figure being somewhere between the two.[3] In 1842, there were 356 machines and 372 vats. While the number of machines continued to rise, there was a steady fall in the vats so that in 1850 there were 412 machines compared with 344 vats.[4] In 1860, hand made paper accounted for only about 4 per cent of the total. By 1870, there were 456 machines and in 1900 over 500 in the whole of Britain but only 104 vats.

Output from the machines was increased in two ways. One was by making a broader sheet of paper on a wider machine and the other was by increasing the speed. Robert's machine produced a sheet of paper just under 24 in. (610 mm.) wide but this became 54 in. (1372 mm.) on the first machine built by Donkin at

Frogmore. The widths of other early machines varied from 4 to 5 feet (1220 to 1525 mm.). 48 in. (1220 mm.) machines were commonly in use in the 1820s and 1830s but in the Scotswood Mill in Northumberland in 1835 there were 'two paper machines of large dimensions, with wires 94 inches broad' (2.36 m.).[5] These must have been exceptional for at the London International Exhibition of 1862, Bertrams exhibited an 80 in. (2.0 m.) machine and Donkin a 90 in. (2.28 m.). Probably more typically, in 1860, the Creams Mill near Bolton boasted three machines of 48 in., 52 in. and 66 in. (1.2, 1.3, and 1.67 m.) width and in 1876, there was a 72 in. (1.82 m.) wide machine.[6] The North Wales Paper Mill at Oakenholt was running with two machines with wire widths of 72 in. and 75 in. (1.82 & 1.9 m.) during the 1890s although, in 1886, Bertrams of Edinburgh had quoted for a machine with a width of 100 in. (2.58 m.).[7] The width would vary, partly on the age of the machine but also on the type of paper that it was anticipated would be made. At Ely Mill, Cardiff in 1894, there were seven newsprint machines with widths of 56 in., 74 in., 79 in., 84 in., 90 in., 96 in. and 106 in. (1.4, 1.88, 2.0, 1.82, 2.28, 2.43, and 2.74 m.). The largest paper machine in the world was started on 7 April 1893 at the Star Mill, Fenniscowles, Lancashire, which produced between 75 and 80 tons of paper 140 in. (3.5 m.) wide each week. This machine cost £15,000 and weighed 370 tons.[8] By the turn of the century, machines 150 and 160 in. (3.8 and 4.06 m.) wide had appeared.

Around 1807, Bryan Donkin was able to raise the speed of his Fourdrinier machine from 20 to 34 ft./min, while that of one of Dickinson's machines in 1817 was only 15 ft./min. George Dickinson was running his Fourdrinier machine at 25 ft./min in 1828 which rose to 40 ft./min. in the 1840s. The machines at the International Exhibition of 1862 achieved 100 ft./min. which was 150 ft./min. in the 1870s. Then there was an astonishing increase so that, by about 1900, 420 to 480 ft./min. had been reached for the fastest newsprint machines compared with 250 to 300 ft./min. about fifteen years previously.[9] By raising the speed, output from the same width of machine could be increased proportionately.

YEAR	NO. MACHINES	MACHINE PROD (tons)	HAND PROD. (tons)
1804	3	10	14,950
1810	17	4,793	14,278
1820	32	8,873	12,675
1830	72	21,313	9,377
1840	191	33,463	9,937
1850	267	57,535	5,426
1860	340	95,971	3,839[12]

A rough estimate gives the production from one vat as 50,000 lbs. (23 tons) per annum. In about 1810, the output from the 760 vats was 16,502 tons and from the

33. The dry end of the same machine with calenders and reel.

six machines 557 tons.[10] It was in 1824 that the output from machines first exceeded that from vats, with 14,459 and 12,750 tons respectively. In 1840, the tonnages were: machine-made 33,463 hand-made 9,937 and in 1860, 95,971 compared with 3,839.[11]

In 1900, while the output from the vats had remained about the same at 3,886 tons, that from machines had risen to 647,764 tons. An early machine might produce 300 tons a year which, in 1851, could be between 5, 6 or even 10 tons a week. This had increased to 1,000 tons a year by 1860 and reached 25 tons per week in 1862.[13] In fact, it was said in 1835 that a machine could make as much paper in as many minutes as formerly it took weeks. In 1838,

159

34. Paper being prepared for despatch at Oakenholt around 1905. Today, with computer control, the stacks of paper would be almost flat.

the three Dickinson's mills had a combined output of 41 tons per week. The output of the largest single manufacturer in 1849 was under 3,500,000 lbs. (1,600 tons) per annum. By 1859, this had increased to five million pounds (2,232 tons) with a total output in the United Kingdom of 97,250 tons.[14] Edward Lloyd Ltd. at the Daily Chronical Mills, Sittingbourne, Kent, was producing 400 tons of paper per week on seven machines in 1895 which rose to 850 tons

160

per week on eleven machines (six of which were over 100 in., 2.58 m.) in 1905.

New Sources of Power

The location of the paper industry had been determined partly by proximity to sources of raw materials, rags, and outlets for the paper, the towns, so that it was situated close to centres of population wherever possible. However the industry also needed power, at first for the beating machinery and then for other types of equipment as more and more of the production processes were mechanised. This need for power could be met at first only by water and very occasionally wind. So the mills had to be situated along rivers where a fall in level could be exploited by building dams, channels and waterwheels. However, water as a source of power was unreliable, due to floods or freeze ups, and, more important as the size of mills grew, it was limited in the amount that could be generated at any particular site.

As more modern technology became available such as iron suspension water-wheels, papermakers availed themselves of the opportunity to increase the potential of their mill sites. The enormous waterwheel now preserved in the Royal Scottish Museum in Edinburgh drove a papermill for a long time. Then for many years into the nineteenth century waterpower seems to have been preferred long after steam engines had proved their reliability, probably because waterwheels drove the machinery more evenly than the early steam engines. John Dickinson built his new Croxley Mill in 1826 and drove it with a waterwheel.[15] Bleachfield Mill at Ayton just north of Berwick on Tweed was built as late as 1842 with a waterwheel. In Scotland in 1872, out of a total of 8,144 h.p. employed in the paper industry, 2,187 still was provided by water.[16] The position was very different in the rest of Britain.

	LANCS.	U.K.
No. of paper mills	42	271
Steam power (hp)	8,368	20,746
Water power (hp)	403	5,473[17]

Some mill owners substituted water turbines for their waterwheels when that technology had improved far enough. The old waterwheels at the Dickinson's Nash Mill were replaced in 1879 with 'Hercules' turbines.[18] This type is included in the list of equipment at Creams Mill when it was up for auction in 1905 but no installation date is given.[19]

However, the enormous growth of the paper industry during the nineteenth century could not have been achieved in Britain with the existing waterpower resources. The answer lay in the steam engine. One of the earliest Boulton and Watt rotative engines, with sun and planet gear instead of a crank, was installed in 1786 at Wilmington Mill near Hull.[20] It was rated at 10 hp. This was very early days for such machines and it is not surprising to find that no more seem to have been installed at paper mills, which had to have copious supplies of water in any case, until Koops's large 80 horse engine in 1801. Steam power was being used in a mill at Chester by 1802 and at Fellingshore in County Durham when the Tyne Steam Engine Paper Mill was destroyed by fire.[21] The first Boulton and Watt steam engine in Scotland was one of 20 hp built in 1803 for the Devanha mill of Brown, Chalmers & Co., at Craigbeg Ferryhill, Aberdeen. This mill failed four years later but the engine was purchased by Lewis Smith of Peterculter who installed the first paper machine in Scotland. In the same year, another steam engine was operating in a mill at Lasswade.[22]

In view of the failure of Koops's enterprise, it is surprising that William Balston should have considered building a mill of such a size as ten vats and in addition powering it by a steam engine in 1805. He gave as his reason,

> Falls of Water, that is Embankments upon streams, may . . . be considered as a species of real security, but no more are they indestructible than the steam engine itself. Both are liable to constant attention and repeated repairs, and in my opinion in this respect the latter has the advantage, so long at least as a supply of fuel can be obtained at a reasonable price. For there are no Falls of water that I have seen but have been erected at great expense, require constant repairs in all instances, and in some most heavy expenses, with additional risk of the attached property in wintry seasons from heavy floods, . . . or to fail altogether, with steam the manufacture could be carried on regularly throughout the year.[23]

In February, Boulton & Watt sent Balston an estimate of £1,698 for a 36 hp engine with a 42 hp boiler which were delivered that May. The engine ran for ninety years.[24]

In 1833, James Wrigley had a 35 hp engine at his Bridge Hall Mills but this was large for that time because, in a survey of the Bolton district, there were six engines with a total of 76 hp, an average of 12.7 hp. Another survey carried out in 1837 by the Manchester Statistical Society, covering a wider area, recorded fourteen steam engines with a total of 312 hp, an average of 22.3.[25] A large beam engine of 40 hp was installed at Apsley Mill in 1845 at a cost of 3,000 guineas but by that time this design was obsolescent. The design of the next ones built there in 1856 showed the latest trends which were influenced by railway locomotive practice. Leonard Stephenson had joined the staff at Apsley Mill and produced a horizontal engine with two cylinders, 12 in. diameter by 24 in. stroke which rotated at the then high speed of 60 to 120 r.p.m.[26] Such engines would have driven the machinery through lineshafting and belt drives.

However, later paper mill engines followed the practice of textile mills, with the names of many famous Lancashire and Yorkshire engine builders appearing on the papermaking scene. It became customary to have a large engine driving the preparatory machinery such as the beaters while the actual paper machine might have its own engine, or engines, one for the wet end and another for the drying cylinders. The engines on the paper machine by the 1840s would be high pressure ones from which the exhaust steam was passed into the drying cylinders. Then for greater fuel economy, engines were made as compounds, with steam being used first at high pressure in one cylinder followed in a second at a lower pressure. By the 1890s, some of the steam from these high pressure cylinders would be bled off and once again used in the paper drying cylinders before being taken back to the low pressure engine cylinder.

In 1882, Galloways of Manchester had supplied one horizontal compound engine to the North Wales Paper Mill which was replaced by another built by Hick Hargreaves, Bolton, just after the turn of the century at a cost of about £3,200. This was a horizontal tandem compound with Corliss valve gear which developed a thousand horse power.[27] These later engines had rope drive around a large flywheel to transmit the power to the lineshafting. Dickinsons had another horizontal tandem compound 450 hp engine supplied by the Yorkshire firm of Pollitt & Wigzell in 1908.[28] This drove the machinery through 5 in. diameter lineshafting. In 1926, the East Lancashire Paper Mill at Radcliffe had Musgraves of Bolton install a pass-out engine of similar layout which developed 2,500 hp. It drove the beaters through line shafting and a d.c. electric dynamo but, at that period, reciprocating engines were becoming obsolete as, in their turn, they faced competition from steam turbines.

During the nineteenth century, the ideal site for a paper mill was determined by a variety of factors. William Balston had chosen a situation by the side of the navigable River Medway where his rags and coal could arrive by barge and the paper be despatched in the same way. John Dickinson's mills were close to the Grand Union Canal which in its turn provided good communications. Later, railways were important, and we find that the Heap Bridge Paper Mills of Yates, Duxbury as well as James Wrigley's Bridge Hall Mills, both near Bury, Lancashire, had railway sidings serving them. New mills, such as the North Wales Paper Company, as well as older ones, such as Fourstones Mill near Hexham, all would have aimed to be rail connected if at all feasible.

Proximity to sources of coal was also important, and all the mills mentioned above were situated in or very close to mining districts. Around 1850, William Joynson said that he burnt four tons of coal for every ton of paper he made. Coal was loaded out of barges on the canal above the East Lancashire Paper Mill directly into the boiler house, while Disley Mill obtained coal from its own mine situated in the hill immediately behind it! With the introduction of esparto and wood pulp, proximity to good harbours became a consideration so that the raw materials did not have to be transhipped far from ocean-going vessels. Therefore the remoter mills in the

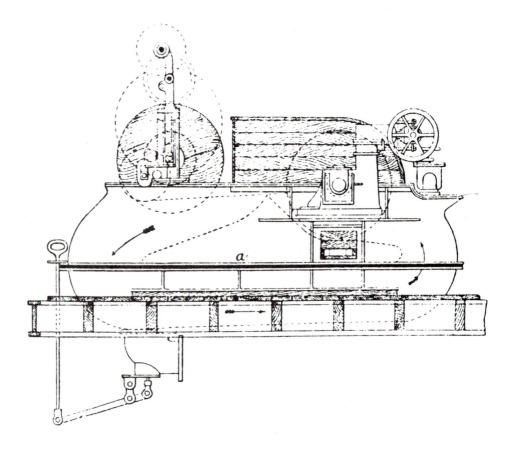

35. Diagram of the Umpherston beater. The pulp circulates back underneath the
roll and bedplate.

smaller valleys closed unless they had some particular advantage, such as producing
a particular variety of special paper or were located near to some source material for
papermaking.

Improvements to Papermaking Machinery

Throughout the nineteenth century, people continually were inventing new ideas to
improve the quality of paper. While the actual production of wood pulp in its various
forms was rarely carried out in this country, converting esparto was and a whole
range of boilers and machines were developed to do this. Some of these have been

mentioned briefly already so that this chapter will concentrate on how the beating engines and the paper machines themselves evolved to enable the vast increases in speeds, widths and production to take place. Once again, the course of the pulp will be followed through the papermaking mill and developments studied in that order and not by date of origination.

The Hollander beater suffered from two draw-backs. The roll performed a dual function, both pushing the pulp round the trough as well as beating it. Then the shape of the early troughs with vertical sides fixed to flat bottoms left pockets

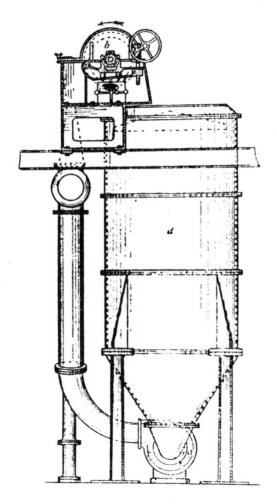

36. The 'Tower' beater. The beating roll is at the top and the circulating pump at the bottom of the tall storage chest.

165

37. The Marshall conical refiner showing the beating bars, c. 1910.

where the pulp might settle and miss being beaten. With the pulp flowing around
the feather in the middle, that on the outside circulated at a different speed from that
at the centre, so not all the fibres were treated equally. The shape of beater troughs
was improved so that the sides and bottoms were curved and the pulp flowed more
evenly without the dead spots. Then radically new layouts were evolved so that all
the pulp would pass equally through the beating bars. Instead of the pulp flowing
to one side of the roll, it passed back underneath in a separate channel. By the end
of the century, two types were popular in esparto mills, the Umpherston and the
'tower' beaters. The Umpherston beater looked similar to the ordinary Hollander
except that the second half of the trough was underneath but the roll still circulated
the pulp.

   The tower beater was much taller with the pulp dropping from the bars of the
roll into a storage cylinder, hence the term tower. At the bottom of the tower, the
pulp was drawn out by a pump which circulated it up to the roll again. It was claimed

that this system needed less power and that the pulp could be beaten more evenly. A.E. Reed patented a beater with a separate circulation impellor in 1892 on which he was able to place the bars much closer round the roll and run it at much higher speed than on ordinary Hollanders. He stated that it mixed the pulp much better with the impellor, beat more of the pulp more evenly because 'on an ordinary beater the great probability is that some of the contents of the engine never get an actual rub between roll and plate at all', 'takes less than half the power consumed by ordinary beaters' and 'no refiners are required'.[29]

So that the pulp could be washed better in the Hollander beater, James Craig patented a form of drum washer in 1839.[30] A cylinder covered with wire mesh was partially immersed in the beater trough so it could be rotated by the flow of the pulp. Clean water was poured in and the mesh restrained the fibres from flowing out with the water which passed through the drum, one end of which was closed and the other fitted closely into a sealing strip to allow the water to flow away down a pipe, rather like a cylinder mould machine. Herring mentioned drum washers in 1855 but they differed from Craig's for in later versions there were curved blades inside them which, as they rotated, lifted the water into a central duct from where it flowed away.

The reference by A.E. Reed to refiners was to those machines which today have virtually replaced Hollander beaters in paper mills. Hollanders, and their different varieties mentioned briefly above, have the disadvantage for the modern papermaker that they produce pulp in batches whereas the capacity of a modern machine needs a continuous supply. The middle years of the nineteenth century saw Americans devising ways of beating pulp in a constant flow. The first of these was Laban C. Stuart who took out an English patent for a disc refiner in 1856.[31] He placed a disc with cutting knives on both its faces between more knives on opposing surfaces on the ends of an enclosed cylinder. The disc was rotated and the pulp to be beaten passed in through the centre of one side of the cylinder, across one face of the disc, round its circumference and back past the other surface. The disc could float on its axis and so probably did not yield a very good pulp. Stuart claimed that the speed of the Hollander was limited, because, if it were rotated too fast, it would not circulate the pulp through centrifugal action and cavitation, that his machine did not suffer from these defects and that it took up much less space. While the disc refiner is used extensively today, its form differs from that patented by Stuart.

The more important machine to begin with was the 'refiner' for which Joseph Jordan took out two English patents in 1860.[32] It consisted of a stationary outer cone with knife blades fixed on its internal surface inside which rotated a cone of slightly different taper, also with knives on its outer surface. The rags were fed in with water at the narrow end of the cones where the tapers were furthest apart and passed out at the larger end where the cones were closer. The knives beat the material as it passed through. In his second patent, Jordan improved the arrangement of the knives so that his refiner has remained practically unaltered up to the present day. It is interesting to note in view of the later introduction of thermomechanical pulp that Jordan proposed to let steam into the beating process. In Britain, 'Jordans' were used particularly for refining the stuff after it had passed through ordinary Hollander beaters to remove any lumps which might have been missed to give the pulp a better consistency, but more and more they have become used to actually beat the pulp, especially when, like mechanical wood pulp, it does not need much fibrillation.

The quality of the pulp was improved also by its being cleaned and having the lumps removed before passing to the paper machine. In his patent of 1817, John Dickinson passed the stuff through a trap, under a large float and then over a sill into the breast-box.[33] In the section before the float, any light articles, such as cork or pieces of wood, would rise up and be taken out. Where the stuff passed under the float, there was a sloping bottom ending in a sink. Here heavy lumps would settle which could be drained off from time to time through a cock. In 1836, in conjunction with William Tyers, he proposed placing a series of magnets underneath a thin copper trough across which a thin stream of pulp flowed to remove any pieces of iron or steel which might cause rust spots in the paper.[34] As this was before the days when electricity was being generated by dynamos, he might have to use between two to four hundred horseshoe magnets. Other ideas for cleaning the pulp consisted of passing it along channels with stepped or ribbed floors where heavier particles would settle, or, in the case of Bryan Donkin, over boards with specially shaped holes. When the holes were full of the heavier dirt, the boards were removed and cleaned.[35]

The problem still remained of how to remove lumps of fibres which had not been beaten properly. If the pulp were passed through any strainer, the holes would soon become clogged unless there were some form of cleaning apparatus. In 1830, Richard Ibbotson proposed using in place of the earlier woven wire strainers a series of flat bars which were mounted inside a container in a horizontal position with fine slots between them so they could be 'joggled' with a quick up and down motion at one end. This helped to pass the fine pulp through the slots. Inside the container, he fitted an agitator to keep the stuff stirred and he removed 'the knots or other matter which from time to time accumulated on the bottom of the said strainer by a narrow hair brush'.[36] As this method of cleaning the slots appears to have been performed by hand, it is doubtful if it were satisfactory. However this may be considered as the ancestor of later flat-bed types.

Other people quickly suggested rival machines. John Hall used a cylindrical framework covered with wire mesh or a coarse laid surface which was meant to work on a principle similar to that of the cylinder mould machine, with the pulp flowing away through it and the lumps retained.[37] How his screen was cleaned is not clear. This was followed quickly by another cylindrical type from John Dickinson constructed like a squirrel cage with very fine slots around the circumference through which the pulp flowed as it rotated like a cylinder on a mould machine. The slots were cleared by a float moved rapidiy up and down inside the cage which drew the correct pulp through and pushed off the thicker lumps.[38] Numerous types have been evolved since which have helped to improve the quality of paper and even have been applied in hand-made mills too.[39]

During the nineteenth century, the stuff flowed out of the breast box onto the moving wire simply by gravity through a 'slice' which always retained primitive forms of adjustment to alter the width of the slot. While in 1828 George Dickinson tried shaking the wire vertically and he claimed that more consistent paper was made,[40] this does not seem to have been taken up by other manufacturers who retained the sideways shake. In 1858, an important advance was made when Thomas Donkin, one of Bryan's sons, patented the wire-guide apparatus (the idea of a Frenchman Gabriel Planche) to keep the wire automatically in the centre of the machine by altering the position of a guiding roller. Up to that time, the screw controlling the position of one end bearing of this roller had been turned by hand but he fitted 'spades', as they are now called, which were pushed to either side if the wire ran off course and which then operated a mechanism to turn the screw on the bearing.[41]

The history of how watermarks were made on machines will be covered in Chapter XII. Here will be examined another important addition to the wire on the Fourdrinier machine. Another part of George Dickinson's patent in 1828 concerned applying a vacuum to draw water out of the sheet of paper while it was still on the machine wire. With his experience of cylinder mould machines, the bottom roller of his wet press he made hollow with perforations. Inside this roller was a vacuum trough to draw out some water. The wire passed through this press and then on to a couching roller at the wet end.[42] This idea does not seem to have come into any general use at that time but reappeared in the form of the vacuum couch in 1897 when it was applied by Alexander Black in the final roll of the wet end.[43] Once again, the idea was not taken up quickly, possibly through the difficulty in forming an adequate vacuum, although today vacuum couch rolls can be found on most paper machines.

It was the vacuum box of James Brown, Esk Mills, Pennycuick, which was quickly applied to Fourdrinier machines, even though he may not have been the first with the idea. In 1836, he intended to make

> a better and more perfect application of a vacuum under the endless wire
> cloth or web of papermaking machines for the purpose of more effectually
> withdrawing the water from the paper pulp as it passes from the pulp vat to
> the couching roller.[44]

He proposed that the box would not have any perforated lid (as possibly others had) and fitted sliding shutters at the sides so the width could be adjusted to that of the sheet of paper being made. It is interesting to note that two vacuum boxes were included in Bertram's specification for the North Wales Paper Company's machine in 1886 but no vacuum couch.[45] In 1839, T.B. Crompton improved the steadiness of the vacuum when, instead of reciprocating pumps, he adapted the centrifugal fan which had been used in opening and scutching machines of the cotton spinning industry.[46] An interesting suggestion for conservation of energy was William Broadbent's patent in 1847 for using the steam which passed out of the drying cylinders to create the vacuum under the wire.[47] It is certainly remarkable

38. Dickinson's disc cutters on one of T. Tait & Sons latest reelers being fed from the widest fine paper machine in Britain. Today, the cutting widths can be selected by computers.

to watch on a modern paper machine how much water is drawn out by the vacuum boxes for it is here that the pulp really becomes consolidated enough to be called a sheet of paper.

As far as drying the paper is concerned, the drying cylinders themselves have altered little over the years. The numbers increased with the speed because the temperature could not be raised for fear of damaging and cockling the paper. In the early days after T.B. Crompton's 1820 patent, the drying cylinders were supplied by manufacturers separately from the wet end, and sometimes might be made by another company. This was a trade in which Robert Stephenson & Co., more usually known for its railway locomotives, engaged for nearly ten years from 1824.[48] Sometime after 1830, the separate felts round each cylinder were changed for a continuous one which might pass over its own drying cylinders too. In 1850, Robert R. Crawford of Fourstones Mill proposed using open nets or porous cloths round the drying cylinders, particularly when drying paper that had been sized.[49] The importance of allowing the steam to escape is recognised today. Instead of a single line, the drying cylinders might be arranged in tiers.

In cutting and reeling the paper, there were a great many ideas which could be fitted either at the ends of the paper machines themselves or on separate machines. One of the most important was included in John Dickinson's 1817 patent. In this he used rotating disc cutters which could slit the sheet of paper along its length as it was being unwound from one reel so that it could be wound up into narrower ones. This principle is the method employed even today.

While improvements were being made to the Fourdrinier machine for making ordinary paper, both it and the cylinder mould machines were adapted for making board. Boards can be divided roughly into two categories, those which have the same type of fibres throughout and those which have a better quality surface, sandwiching a cheaper, inferior middle. The intermittent board machine must have made its appearance in the 1830s for it was certainly being used at the Bleachfield Mill in 1842. At the Schoolmaster windmill in the Zaan district, Holland, the principle can be seen applied to a Fourdrinier machine installed there in 1877. At Schieder Mill near Hameln, West Germany, an ancient cylinder mould machine is employed. The latter type is still being used at B. & S. Whiteley's Pool Paper Mill in Wharfedale. A sheet of paper is formed in the normal way on either type of machine but is wound around a roller under pressure which causes the layers to stick together. When sufficient thickness has been built up, a bell rings and the operator cuts the sheet by drawing a knife along a slot in the roller and pulls it off. Only board with the same material throughout can be made on such machines. That made at Aytoun, the Schoolmaster and Scheider mills was dried in lofts like hand made paper.

For longer boards, the cylinder mould machine was developed at first. In 1817, John Dickinson designed a machine for spreading size or paste on either two or four rolls of paper which would cohere when passed between pressing rollers. Of greater significance for subsequent development of later board machines was his proposal at the same time for making a two-ply paper on his cylinder mould machine. A sheet of paper would be made first and reeled up while still wet. This reel was suspended above the vat of preferably a second machine which would have a different type of pulp. The sheet from this second vat would be combined with the one from the reel in the wet press which 'united them into one body'.[50] Dickinson realised the advantages of this system when he wrote in his patent,

> By the process above described, in papers where only one side is required for use, such as principally drawing and plate papers, that side may be produced of finer colour or of different colour, and of finer and closer texture, and at less expense than would be possible if the whole mass were the same.

These principles remain true for making boards even today.

It was not until 1831 that John Dickinson patented the next stage in the evolution of the board machine when he proposed having twin vats, similar to those already in use.

39. The air-drying machine at Sawston mill, near Cambridge, in 1974.

> The sole novelty consists in bringing the paper first formed [in one vat]
> into contact with the paper formed on the other cylinder previous to its being
> detached from the felt which it is first made to adhere to, so that being in a
> very wet state the two layers of paper unite better than if either of them had
> been first pressed and detached from the felt.[51]

It was left to the Americans to turn this into the multi-vat board machine with
a line of vats which could be filled with pulps of different varieties so that board
with a good quality surface on either side could be produced with layers of inferior
quality in between. In 1863, J.F. Jones, of Rochester, New York, took out an English
patent for a machine with seven vats for making heavy boards. The combined board
passed over steam drying cylinders arranged on a floor above the vats. Towards the
end of the century two multi-vat machines were installed in this country, one at
Cannon's Mill, Sandford-on-Thames, near Oxford and the other at Hedley's Mill,
Loudwater.[52]

If the paper were to be used for printing or writing, it needed to be
treated to reduce its absorbency. Since some time early in the eighteenth century,
aluminium sulphate, know as 'alum' to the papermaker, had been added to gelatine
to produce a harder sizing which, upon drying, became less soluble in water than
gelatine sizing without it. Unfortunately, with time, alum could produce sulphuric
acid which caused both discolouration and also degradation of the paper. The same
type of paper might contain different amounts of alum depending upon the weather
because papermakers increased the amount during hot days.[53] In 1807, M.F. Illig
published his pamphlet of his experiments with rosin sizing which he had carried
out seven years previously.[54] Rosin (colophony) would combine with alum to deposit
a hydrophobic precipitate, or form of size, which was spread evenly over the paper
fibres by the application of heat. This could be added during the beating stage and
was know as 'engine sizing'. Although not widely used in hand vats, it was adopted
in the 1830s when drying cylinders were installed on paper machines because the
paper left them already sized. However the acids formed in the process caused the
paper to deteriorate later. Another problem was that the felts could become clogged
very quickly unless they were properly washed.[55]

Gelatine sizing on the surface of the paper still was employed for the better
quality papers. On a machine, either the continuous sheet of paper could be
totally immersed in a vat of gelatine or the gelatine could be applied by rollers.
In yet another section of his 1817 patent, John Dickinson proposed sizing rollers
for putting on a layer of size which he found needed to be stronger than normal.
Strangely, although the paper had just passed over drying cylinders before sizing,
he did not try to dry it on the machine after it had been sized. A drawing of a
Fourdrinier machine supplied to Matthew Towgood in 1830 shows the sheet of
paper passing round three rollers in a tub of size, the first time this was tried. Excess

size was taken off by a doctor blade. The paper continued over drying cylinders afterwards.[56]

The air drying machine was developed by R.G. Ranson of Ipswich and S. Millbourn in 1839. After passing through a tub-sizing vat, the paper was taken over a serious of skeleton drums inside which paddles rotated to direct hot air against the paper.[57] Various forms of this machine appeared. One is described in Robert Crawford's patent in 1850[58] and machines were built with 40 or 50 drums. Then, in the 1870s, the Wrigleys at Bridge Hall Mills were investigating various types without much success.[59] In the Busbridge mill at East Malling in Kent in its final layout in 1930, the paper passed from the wet end of the machine, over steam drying cylinders on a higher level and through a size bath. It then went along the whole of the top floor of the mill round 36 air drying drums before descending to the bottom floor, through a calender stack for glazing and finally past a slitter and cutter.[60] To regulate all the different sections of this machine at the correct rate must have been a work of art. Twogoods at their Sawston Mill near Cambridge had a separate section for the sizing and air drying.

The people who first introduced the paper machines were stationers who were interested in selling paper of high quality. The new machines certainly reduced the price of paper but it is difficult to know what quality they produced. In 1829 a contemporary remarked,

> About twenty-five years ago these machines were introduced, and they form another era in the deterioration of the paper manufacture, for by means of them an immense quantity of inferior paper was thrown onto the market. The machine required, at least in its original construction, that the pulp should be ground to the finest possible consistency, and being necessarily void of fibrous matter, its cohesive strength was thereby reduced.[61]

Yet only four years later, John Dickinson could write to Charles Longman about the Scottish manufacturers,

> I arrived here this evening and left all friends in the North favourably dispos'd, but the makers there are really turning out capital paper, better than ever I saw before.[62]

Probably both comments are true, for while poor quality paper can be turned out on any machine, standards were raised quickly during the nineteenth century as a result of all the new inventions throughout the whole papermaking process.

The mechanisation of the paper industry gave rise to a new industry in its own right, that of building papermaking machinery. There were people who specialised in papermaking equipment even before the days of the machine. John Hall founded his Dartford works in 1785 and installed and repaired equipment in the Fourdrinier and other mills. Donkin was apprenticed to Hall and left to start a mould making business at Dartford in 1792. John Marshall was apprenticed to Donkin and took this over when Donkin began to work on the Robert machine for the Fourdriniers. Marshall later made paper machines as well as dandy rolls.[63] Donkin established the first works for building paper machines and by 1851 his company had constructed 191 machines, of which 83 remained in Britain, 23 had been exported to France, 46 to Germany, 22 to northern Europe, 14 to Italy and southern Europe, 2 to America and 1 to India.[64] Exports have always remained an important part of the machinery business, and some of the earliest work on the paper machinery side carried out by Robert Stephenson & Co. was sent in 1824 to Ireland.[65]

As the number of paper machines increased, so did the numbers of paper machinery manufacturers. William Bertram, a millwright and engineer apprenticed to John Hall, founded the famous works at Sciennes, Edinburgh, in 1821 where he was later joined by his brother George. George Tidcombe started to build machines at Watford to the north of London in 1827. In fact, most centres of papermaking soon had their own machinery makers. Specialist firms appeared. For example, James Kenyon, a weaver of Bury, was making felts for paper machines by 1827, if not earlier.[66] John Dickinson was weaving his own wire cloth at Apsley Mill in 1822 while in Scotland, William McMurray set up a specialist firm for this in about 1837.[67] In this way, developments in one industry created openings for new enterprises so the paper industry came to need a whole range of supporting manufacturers, just as it served other businesses with the paper it produced.

XII

WATERMARKS ON PAPERMACHINES

Whatever type of wire N.L. Robert may have used on his machine, the Fourdriniers and John Gamble definitely mention wove wire in their patent of 1807.[1] In fact, ever since this date, the main machine wire on all subsequent Fourdrinier machines has been made on the wove principle and therefore, on its own, will give no watermark. While there have been suggestions for attaching wire profiles on the machine wire, the problem of securing them as the wire passed round the end rollers is similar to the problems facing the first inventors of watermarks in Italy when paper was formed on moulds with removable covers. The constant flexing of the wires would have caused the wire profiles to quickly fall off. To have attached enough wire profiles on the length and breadth of a Fourdrinier wire would have needed a great many identical wire profiles which were not available early in the nineteenth century even if the idea had been practical.

People did try to place wire profiles on the machine wire. For example, in 1826 L. Aubrey patented an idea for weaving the watermark into the wire.[2] Four years later in 1830, Thomas Barratt, a papermaker of St. Mary Cray, Kent, tried to sew watermarks onto the machine wire in the same way as on hand moulds. Also he placed very small rollers under the wire which he claimed helped to remove the water more quickly. One of the disadvantages of his system was that the complete machine wire had to be changed whenever the watermark had to be changed. Also the length of the wire had to vary to accommodate sheets of different lengths. There is no evidence to show that it was ever a commercial success.[3]

John Dickinson's first patent in 1809 explains how the cylinder of his machine could be covered with

> a connected web of laid wire, which may be laid over the cylinder exactly in the same manner as the wove wire, observing that the laid wires should be parallel with the axis, and the tying wires at right angles with it, observing that the laid wires must be very fine, placed very near together, and drawn as tight as possible at the ends.[4]

However, no paper seems ever to have been made at this period on this type of cover as far as we know.

176

In 1823 and 1824, both John and George Dickinson were looking again at making laid paper on the cylinder machine.[5] A book, *The Century of Inventions*, written originally by the Marquis of Worcester in 1655 but printed by C.F. Partington in 1825, has paper in which distinct laid lines appear but no chain lines. It is possible that this may be some of Dickinson's experimental laid paper. At the present moment, nothing is known about who had the idea, or when, of sewing wire profiles onto the surface of the cylinder mould machine. Portals installed one in 1917 for producing paper for one pound notes when it was described as 'not long been invented'.[6] The smaller surface of the cylinder meant that fewer wire profiles were needed than for the wire on a Fourdrinier machine. The surface of the cylinder never flexed and so did not present the same problems as the moving wire of the Fourdrinier machine.

The way in which the fibres are deposited on the cylinder of a mould machine is not too dissimilar from making paper by hand. The fibres form themselves around the wire profiles and eventually cover them so that a watermark with almost the same clarity can be produced.[7] Cylinders may be covered with either wove or laid wire and both Hodgkinson's at Wookey Hole and W. Balston's at Maidstone had detachable covers so that a change could be made quickly from say a Whatman laid to a Whatman wove paper. By the 1850s, many identical copies of a watermark could be made by the electrotyping process which could be sewn or soldered onto the surface of the cylinder. Then a woven wire cover was not too large to have a

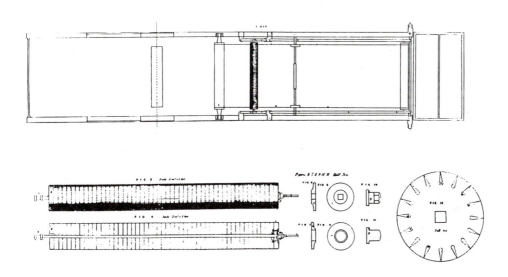

40. J. & C. Phipps' patent for a dandy roll, 1825.

multiplicity of three-dimensional watermarks pressed into it. Today, the majority of high security paper for bank notes and the like, with portrait pictures and other complex watermarks, is made on cylinder mould machines.

The problem of how to watermark paper on Fourdrinier machines was solved in two stages. The objective of John and Christopher Phipps, who had offices in Upper Thames Street, London, and two mills at Dover, River Mill and Crabble Mill,[8] was to make laid paper on the Fourdrinier machine and in 1825 they patented a 'roller, the superficies or cylindrical part of which is formed of laid wire'.[9] This roller was placed towards the end of the machine wire where most of the water had drained out of the pulp. It was constructed around a central axle which was held in position by bearings on either side and was placed at right angles across the machine. Its surface rested on the pulp so it revolved as the pulp on the machine wire passed below it. As it rotated, it impressed whatever patterns had been made on its surface into the pulp. Subsequent pressing to remove the water and drying did not eliminate these identations so they remained in the paper as watermarks.

The description of these 'dandy rolls', as they were called later, given in the patent is so thorough that the Phipps must have carried out experiments with them before submitting their patent specification. The cover could be constructed in two different ways so that the laid lines ran either across the width of the sheet or along its length. The axle was the width of the machine and had circular wooden ends placed a little wider than the width of the sheet of paper. If the laid lines were to run the length of the sheet, then ribs like those of a hand mould were stretched across the dandy roll between the two ends at intervals that were the same as those on a hand mould. The laid lines were formed by taking a long length of wire and winding it around the circumference of the dandy roll in a tight spiral. At each rib, a pair of wires were twisted round this laid line to create the chain lines which were bound to the ribs.

If the laid lines were to run across the width of the sheet, a laid cover, a little wider than the sheet, would be made and wrapped around the circumference of the dandy roll. The chain lines were thin metal discs which acted as spacers and strengtheners to which the cover was sewn. The patent suggests that these discs were not attached to the main axle as they were in later rolls where they gave more rigid support. The Phipps had tried dandy rolls of various diameters from five to twelve inches and had not determined any preference except that thicker sheets needed a heavier impression.

John Marshall is said to have made a dandy roll in 1826 which has been claimed as the first.[10] If this dandy roll is the one now preserved in the National Paper Museum in Manchester, its construction is similar to those described in the Phipps's patent. The answer may possibly be that the Phipps had some experimental dandy rolls constructed first for their own use possibly by Marshall who then built the first one for sale to an outside customer in 1826. Dandy rolls did not vary much in their construction until a few years ago. The wood and iron of the early ones was later

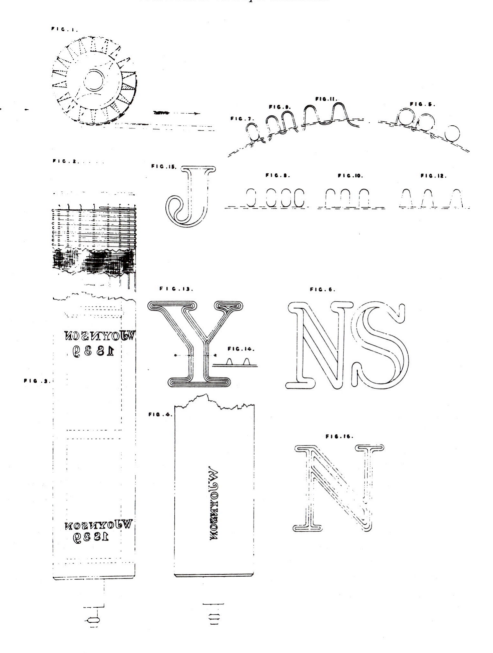

41. W. Joynson's patent for making watermarks with dandy rolls, 1839.

replaced by bronze and the standard cover became the second type with the laid lines running across the sheet.

It was soon realised that dandy rolls improved sheet formation and even assisted in dewatering. Quickly dandy rolls appeared with wove covers which gave these advantages to wove paper too. John Marshall was making wove dandy rolls as early as 1828.[11] In fact, these other advantages of the dandy roll have assumed almost more importance over the years than the watermarking function. Three years after the Phipps' patent, J. Palmer, papermaker of Mile End, London, patented 'wire rolls to assist in forwarding the manufacture of paper by pressing out part of the water before the pulp passes under the wet press cylinders'.[12] His device looked remarkably like a dandy roll and occupied the same position on a papermachine. An illustration, dated to 1830, of Towgood and Smith's paper machine shows two 'hollow wire rollers which press slightly on the newly formed paper while it remains on the wire to press out some water'.[13] Then in 1830, John Wilks, a partner in Bryan Donkin's company, had a sort of dandy roll also running on top of the sheet of paper on a Fourdrinier machine with a vacuum suction inside it to draw off some of the water upwards.[14] It is interesting to note that attempts at dewatering the sheet on its upper surface were being tried as early as this.

While the dandy rolls could impress watermarks of a sort, the chain and laid lines, into the slushy paper, they still lacked the capability of forming full watermarks. In 1839, William Joynson patented 'affixing letters, figures or devices upon a revolving axis . . . known by the name of the dandy roll, the dancer, the top roller'.[15] Just as the Phipps' laid and chain lines were impressed into the pulp, so were Joynson's devices, leaving watermarks. While round wire could be used, in order to stop fibres catching underneath, Joynson recommended 'metal wire . . . with rounded tops and flattened sides and flat bottoms . . . or best of all, with rounded tops and tapered sides and flat bottoms'.[16] Once the principle of sewing wire profiles onto dandy rolls had been established, all the later types of watermarking devices, such as electrotypes and embossed wove wire designs, became applied in their turn. William Frederick de la Rue adapted the three-dimensional or shadow watermark to dandy rolls in 1869[17] so that watermarks of most descriptions could be produced on the Fourdrinier machine.

Yet, because the sheet of paper had been already partly formed on the machine wire and the dandy roll merely impressed the wire profile into it, these watermarks rarely achieved the clarity of ones made by hand or on the cylinder mould machine. Therefore there have been many experiments to improve not only the watermarking aspects of dandy rolls but also to enhance the other qualities they can give to a sheet of paper. A correctly installed dandy roll will help to close the sheet, improve the look-through, strengthen the paper and reduce pinholes, as well as assisting drainage and smoothing the top surface.

Investigations have shown that Phipps was wrong when he said that he found no preference for rolls of greater diameter because in fact a large one is more effective at the same machine speed than a small one. This is because the larger the diameter, the longer will the roll be in contact with the fibres on the wire, giving them more time to change position and reconsolidate. Also the more gentle will be the build up of pressure on the mat of fibres so that the fibres can rearrange themselves more easily. There must be sufficient water still left in the slurry on the machine wire for the fibres to be able to change position but, once the sheet has been formed with the watermark, the fibres must not move again. All this must happen before the dandy roll loses contact with the paper.

At first, the diameter of the dandy roll was often determined by the size of the sheet of paper. To reduce the number of wire profiles needed, the circumference of the dandy roll was the same as the length of the finished sheet of paper, so there was only one watermark (or series of them across the roll) for each revolution. However, as the speed of paper machines increased, the speed of the dandy roll had to increase and friction in its bearings caused it to drag on the pulp. Also the larger the diameter of the roll, the less was the tendency for it to throw drops of water onto the sheet. So it was found preferable to increase the diameter of the roll which, while the circumferential speed had to remain the same, reduced spots from water splashes and breakages in the web. Also, as the width of paper machines grew, so the dandy rolls had to be increased in diameter to make them more rigid. But, as the diameter and length of the rolls were enlarged, so they gained weight and became too heavy to operate without severe drag.

Today, no longer do dandy rolls just sit on the top of the sheet of slurry or paper as it passes beneath them. They are mounted on elaborate brackets with air or hydraulic lifting mechanisms so that they can be lowered to a precise distance above the machine wire to give the correct depth of penetration and also be lowered only when the machine and the dandy roll have reached their proper running speeds. Then drag has been eliminated by driving the rolls by their own motors. On a Fourdrinier machine, the web of paper stretches in the machine direction through tension around the drying cylinders and contracts across the width as it dries. So the dandy rolls can be driven faster than the machine wire to compensate for the subsequent contraction and this minimises the amount of distortion necessary in the watermark.

The firm of W. Green Son & Waite has made important contributions to the design of modern dandy rolls. One is the 'Plate Laid' in which the laid wires are not bound together with chain lines but instead pass through holes around the circumference of a metal disc. This disc can form part of the structure of the dandy roll itself, giving a much stronger construction to both the roll and also the cover. While developed many years before the advent of the high-speed paper machine, plate laid rolls have proved invaluable because they can be run more quickly than rolls with conventional covers. With them, less water is picked up and thrown over the sheet of paper and also they are less prone to trap fibres in the weave of the laid cover, a cause of blemishes in the paper.

Then in 1965, Cecil Douglas and Kenneth Senior introduced a new concept in

42. Pulling the cover on a Dougsen dandy roll.

dandy roll construction with the invention of the Dougsen Dandy. This utilised a
sort of space frame construction with an open wire cylinder replacing the conventional plate supports. The Dougsen roll gives vastly improved drainage, eliminates
the problem of water collecting and being thrown off to cause spots on the paper,
and enables machines to be run at significantly higher speeds. The construction
enables stiffer but lighter rolls to be built, which can have their surfaces precision
ground for greater accuracy. They have been built from 10 in. to 4 ft. diameter (254
mm. to 1.2 m.) and from 7 ft. to 30 ft. long (2.1 to 9.1 m.). The same principle has
been applied to cylinders for cylinder mould machines. Being formed from welded
stainless steel, they are corrosion free and last a long time.[18]

With the continued increase in machine speeds and the need for longer periods of running without changing the dandy rolls for cleaning or other maintenance, it became necessary to fit showers to wash the covers. With the Dougsen type of construction which is hollow inside without any spokes or ribs, the dandy roll can be fitted with sprays and cleaning equipment through its interior. First of all, a steam

43. A Dougsen dandy roll with electrotype watermarks showing the tightening ring at the end.

183

shower can be placed in the bottom of the roll just beyond the nip. Here it helps to reduce the tendency of the stock to be lifted by the roll and so reduces air bubbles and it also assists in cleaning. Then an oscillating internal cleaning shower operated with water at high pressure can be fitted inside. This reduces time spent on roll changes. It is worked in conjunction with a splash guard that catches all the shower water and any other water that is thrown off the roll, and diverts it away from the web. The drive can be imparted at the circumference of the dandy roll where there may be other rollers to support the weight. Today, a dandy roll with a plain wove cover can be run at 500 m/min machine speed rotating at up to 140 r.p.m.

Finally a vacuum suction box has been fitted underneath the machine wire below the dandy roll. This is in two parts, a water box and then the vacuum box. The wire passes over the water box which prevents further drainage of water so that the web is floating on the Fourdrinier wire just prior to the nip of the dandy roll. The fibres forming the web are therefore at rest as they enter the nip of the dandy roll. The vacuum box is situated just beyond the maximum nip of the dandy roll where the vacuum can remove excess water from the sheet and fix the fibres in their relatively undisturbed position while the dandy roll closes the sheet on the top side. By drawing the web away from the surface of the roll, the vacuum chamber minimises the problems caused by turbulence such as air bubbles and the stock lifting. All of these improvements have helped to give watermarks made by a dandy roll on a Fourdrinier machine much greater clarity than they ever had before.

XIII

THE TWENTIETH CENTURY

The Background

The last thirty years of the twentieth century have seen sweeping changes in the British paper industry. While it might be said that the basic principles of the way paper is made have not changed from the nineteenth century, there has been profound scientific study into what is happening at the various stages so that today papermaking has become highly technical to ensure maximum utilisation of resources, whether of raw materials, the use of energy, recycling chemicals or the prevention of pollution.

The population of the country has risen from 45.4 million in 1907 to 50.2 million in 1950 and remained virtually static at 55.7 million from 1970 to 1977. But demand

YEAR	MILLS	MACHINES	MANPOWER
1972	146	410	63,038
1973	144	396	62,415
1974	143	398	64,741
1975	140	402	62,116
1976	138	358	57,418
1977	139	335	58,812
1978	135	342	59,293
1979	132	325	56,074
1980	123	293	49,290
1981	114	270	43,461
1982	98	253	39,656
1983	88	214	34,930
1984	86	198	34,337
1985	89	195	31,949
1986	89	192	31,860[2]

for paper has continued to increase from about 30 lbs. (29.1 kg.) per head in 1907, 250 lb. (128.8 kg.) in 1970 and 322.7 lb. (146.7 kg.) in 1986. This is much less than consumption in Japan, West Germany, Canada, Sweden and the United States of

America for in the last of these countries the figure had reached almost 300 kgs. per head in 1986. Output per mill has risen dramatically but the size of the industry has contracted through import penetration which has led to a steady decline in the number of mills. In 1902, there were 229 mills with 540 machines in England and Wales with around 60 mills in Scotland. In 1939 these figures had shrunk to a total of 255 mills which had decreased to 228 in 1943. By 1970, the number had fallen to less than two hundred, standing at 195.[1]

The figures quoted in the table above must be seen against the background of production. A great increase in productivity has been achieved in fewer mills with a smaller number of employees. In 1938, output was 2,541,000 tons when 66,870 people were employed. The peak in numbers of employees was in 1959 at 101,000 when output was 3,647,000 tonnes. The highest tonnage was in 1969 at 4,933,000 tonnes with only 93,000 people. When the latest production figures are compared with the numbers of those employed, it will be seen that productivity has continued to rise.

These figures show how domestic production was outstripped by imports for the

YEAR	APPARENT CONSUMPTION '000 tonnes	PRODUCTION '000 tonnes	IMPORTS '000 tonnes
1858	122	87	41
1875	220	162	61
1885	691	369	340
1895	1,027	543	543
1907	1,321	887	451
1912	1,524	1,085	508
1924	1,788	1,317	713
1930	2,506	1,691	1,054
1935	3,174	2,286	1,086
1938	3,596	2,541	1,046
1945	–	1,322	344
1950	–	2,610	680
1955	4,140	3,297	1,105
1960	5,312	4,064	1,426
1965	6,112	4,537	1,723
1970	7,179	4,903	2,506
1975	6,017	3,616	2,645
1980	6,837.0	3,793.4	3,509.7
1981	6,986.6	3,379.7	3,908.9
1982	6,750.0	3,197.6	3,951.4
1983	7,159.0	3,297.7	4,284.8
1984	7,595.6	3,591.0	4,527.9
1985	7,711.0	3,681.2	4,604.1
1986	8,068.2	3,941.2	4,757.1[3]

first time in 1981 but today the position is beginning to be reversed with the new machines which are being built coming on stream. The figures also show the difficult period through which the paper industry, along with others had been passing, with a period of depressed demand in the 1970s as well as changes in trading regulations.

Recent government policies have affected industry dramatically. In the earlier part of this book, we have seen how the British paper industry grew to supply most of the domestic consumption by 1800. Tariff walls gave some protection up to 1861 when all duties and taxes on paper were abolished. Although this opened the British market to foreign competition, production within this country continued to expand and meet about two thirds of the demand. This was due probably to technological lead, a favourable situation regarding power supplies with coal, and the expense of shipping paper from abroad. It is strange that the British paper industry never developed a large export market in the same way as the cotton industry for, with the exception of some speciality papers, exports of 'raw' paper remained comparatively small, although a great deal was sent abroad in the form of packaging, wrapping paper, books and the like.

It may be said that up to the First World War, the industry was generally prosperous through the strength of home demand and also preference in the markets of the Empire. One problem about which it has always complained has been that, while 'free trade' was practised at home, paper industries in other countries were protected in their own home markets and so could subsidize their exports to Britain. The early years of this century saw an increase in import penetration. The increased demand that resulted from the Boer War in 1900 led to an expansion in imports of about 37.6 thousand tonnes, but this trend continued after the end of the war in 1901. In 1902, imports were 43 per cent of the home production figure of about 762 thousand tonnes and about 32 per cent of home consumption.[4]

At the start of the First World War, Britain was the third most important paper producer, with the United States in the lead, followed by Germany which had outstripped Britain by 1905. After that war, the German industry was equipped with new machines, which gave it a technical lead over its competitors and so it could expand its exports further. In 1916, a Royal Commission was set up to ensure supplies for war-time needs and to economise imports of paper-making materials. Therefore the British mills came under government control and their importance to the national economy began to be realised. When competition became acute in the 1920s, the industry was protected to a certain extent by duties imposed under the 1921 Safeguarding of Industries Act, and a duty of 16⅔ per cent *ad valorem* was imposed on all imports of wrapping and packing grades in 1926 for a five year period. More wide ranging duties followed under the Import Duties Act of 1932 which gave a measure of protection to the industry up to the Second World War. Exports to the Empire during this period did not resume their previous levels and so failed to make up for import penetration. However the industry did modernise itself during this period for in 1939, fewer mills were producing about three times the output of 1902.

During the Second World War, the paper industry once again was taken under government control. Once again its importance to the country was recognised for, to design a battleship, twenty tons of paper might be required. Until April 1950, all imports of wood pulp, esparto grass and pulp wood (i.e. wood for pulping) were made through the Paper Controller and it was not until 1956 that the last of the government controls on imports of raw materials and paper was removed. In this post-war period, the British industry faced difficulties in securing new machines because most engineering production was earmarked by the government for export to earn foreign currency, but of course this gave foreign papermakers an advantage through being equipped, once more, with the latest technology. However, price controls gave a stable framework on which it was possible to plan new investment and there was a massive re-equipment during the 1950s when new machinery did become available.

> In the 1950s the British industry got into its stride after the dislocations of war and the strain of post-war efforts to recover. The year 1959 was the last before the E.F.T.A. Treaty was put into effect, and it was a year of record production – 3,647 thousand tonnes compared with 2,541 thousand tonnes in 1938.[5]

In 1960, the industry's prospects changed abruptly with the implementation of the European Free Trade Association Treaty for, under its terms, Norway, Sweden, Finland and Austria, the main producers of pulp and paper in E.F.T.A., were to have their imports admitted with a yearly 10 per cent reduction in duties until 1970 when all duties would be terminated. In the event, the process was accelerated and all imports became duty free at the end of 1966. This exposed the paper industry to the competition of countries which had extensive supplies of raw materials in their woods where integrated pulp and paper mills were beginning to be established. Such mills could produce paper more cheaply than could those in Britain which had to buy their pulp from these same countries. By 1969, the industry appeared to have adjusted to the changed import situation within the E.F.T.A., helped by an upswing in the British economy generally and so an increase in consumption. A record output of 4,933 thousand tonnes was achieved in 1969, a peak year, followed by a relatively good year in 1970, with 4,903 thousand tonnes.[6]

Entry into the European Economic Community in 1973 once again changed prospects for the paper industry with competition coming from new sources as tariff barriers with other European countries were reduced. Then in 1974 came the huge boost in world oil prices through war in the Middle East which has been described as the end of an era of low cost energy. Subsequent inflation, a drop in consumer confidence and a slowdown of industrial production created an economic tailspin from which we are only just beginning to recover fourteen years later. This led to a period of low consumption, slow economic growth and modest increases in demand for pulp and paper products. However, recently confidence has returned again and massive new investment is being made in the paper industry. This is concentrating on resources available locally such as waste paper reclamation and the utilization of such forest resources as we have. The graphs showing import penetration which had risen steeply from 50 per cent in 1980 to 60 per cent in 1983 now reveal a turn down to 59 per cent and this should improve with the commissioning of new machines in the near future.[7]

Energy forms a considerable proportion of the total costs of paper production through the enormous amounts consumed in the drying process and so access to a cheap source always has been essential for papermakers to remain competitive. In addition to the external factors affecting demand for paper, Britain's position of being an industrial leader through its favourable source of energy, e.g. coal, was gradually eroded during the twentieth century in two ways. First the development of another source of fuel, oil, has lessened the importance of coal throughout the world and, until the discovery of oil and gas in the North Sea, Britain equally was dependent as most other countries on imported oil and so had no price advantage in that commodity. Then generation of electricity on a large scale in central power stations has made its production at a mill less economic. Today electrical power is available at virtually any place in Britain at the same price through the development of the national grid distribution system in the 1920s and there is no longer any advantage in a mill being situated on top of a coal mine. Electricity can be generated by gas, coal, oil, water and more recently by nuclear systems so that the price of energy could be nearly the same in countries all over the world except those endowed with special natural resources. Mountainous countries where the pine trees grow for making paper may have the possibility of generating cheap hydro-electricity but in Britain recent government policy has been to keep the price of power artificially high which has been a disadvantage to the paper industry.

The steam turbine replaced the reciprocating engine during the early part of this period because it was inherently more efficient and was more suitable for running continuously. Sometimes these might drive existing lineshafting but more generally they were connected to electric generators so that the current could be distributed to motors in different parts of the mill. In 1915, Bridge Hall Mills installed a 2,000 hp turbine powering a 1,500 kilowatt A.C. generator supplied by British Westinghouse Company of Manchester.[8] As in the reciprocating engine, steam could be taken out of the turbine at an intermediate pressure stage and used in processing machinery around the mill or in the paper machine drying cylinders, but the problem was how to balance the different mechanical and heating loads to conserve as much energy as possible. Steam is still necessary for some stages in the papermaking process such as the drying cylinders and it may be raised by burning waste products particularly from the pulping processes. This was another advantage that foreign competitors had in their pulp mills for they had a cheap source of fuel in waste wood such as bark or from the residues in the sulphate treatment process.

On the one hand, modern technical developments have reduced steadily the amount of energy or heat needed to produce paper. We have seen that in about 1850, four tons of coal were needed to make one ton of paper. Forty years ago this had been reduced to one ton of coal. This figure has been halved within the last six years through better utilization of heat and better machines. On the other hand, energy costs have risen so that, from being about 3 per cent of manufacturing costs in 1973, they have risen to a current level of 15 per cent. Steam turbines have become uneconomic and mills were switching to purchasing electrical power but now this trend is being reversed through the availability of natural gas and the development of more efficient industrial gas turbines.

In the gas turbine, natural gas or light oil is burnt at high pressure in the turbine which is used to generate electricity. The exhaust gasses at a temperature approaching 500° C. pass through a waste heat boiler to raise steam. If insufficient steam is generated, supplementary fuel such as bark can be fired in this boiler. Efficiency of these systems can be 75 per cent which is higher than in a large electricity generating station. The efficiency can be increased by installing a back pressure steam turbine to use steam from the boiler which is then passed to the steam drying cylinders. The steam turbine also generates electricity. The first gas turbine was installed in a Birmingham paper mill in 1985.[9] Another example is the gas turbines which started running in 1987 at the Purfleet Board Mills and generate 10 megawatts of electricity from methane gas derived from a near-by rubbish tip.[10]

Water Treatment

The paper industry is one of the biggest and most efficient users of water in the country. Sometimes treatment is necessary before the water can be used in the mill, either to remove hardness which might scale up the boilers or to prevent contamination from dirt. In the 1950s, Yates Duxbury had to install a treatment plant capable of purifying two and a half million gallons every day, a volume much greater than other plants could deal with at that time.[11] Wherever possible, supplies of clean water are sought to minimise the expense of treatment.

Wolvercote Mill near Oxford and the Dickinson mills on the river Gade were causing concern about the pollution they were causing as early as the 1860s. Legislation has brought stricter controls over the disposal of effluents and standards of purity for discharges were laid down in the 1912 Royal Commission. These were defined further in the Control of Pollution Act, 1974. At the Dickinson's Nash Mill, a Swedish treatment plant was installed in 1934 to precipitate any fibres and size before the water was returned to the stream.[12] Today recycling plants

44. Section through a disc refiner, c. 1900.

have been installed in virtually every mill and when the water is finally returned to the river, it is cleaner than when it was taken out. Portals, the bank note manufacturers, developed such excellent apparatus for purifying water to avoid polluting the trout streams of Hampshire, that a subsidiary company was set up to produce it for sale to others.

Pulp Preparation

Since the Second World War, the traditional raw material, rags, virtually has disappeared from the papermaking scene through pure linen and cotton fabrics being replaced by ones containing man-made fibres which cannot be beaten to form hydrogen bonding. Instead cotton linters, the very short fibres left over from the spinning operations, have become a substitute in papers with a rag content but these form only a small tonnage. With this change in fibres, there has

191

45. Papermachine No. 6 at New Thames mill. On the left can be seen the pressure former delivering the pulp to the wire. This machine underwent a £15 million rebuild in 1984 / 5 putting Bowaters at the forefront of the European production of uncoated woodfree paper.

been a change in beating apparatus because most wood pulps and certainly waste paper do not need to have their fibres shortened but only need fibrillating and the clumps of fibres dispersing.

Not only has batch production in Hollander beaters ceased in all except small speciality mills but the conical refiner is fast disappearing too. One reason for discarding the Jordan is the cost of making the shell and the plug as well as maintenance time to change the bars. Also Jordans both cut and

46. The electrically heated slice lip adjusters controlled by computer on James Cropper's
No. 4 machine.

fibrillate the stock but cutting is their outstanding feature, which is not wanted
with the pulps of today.[13] They are being replaced by disc refiners in which
one plate with blades on it is held stationary while another with more blades is
rotated close to it. Sometimes both discs rotate in opposite directions. The pulp
passes from the centre to the periphery and the gap between the discs can be
controlled with micrometer accuracy and the fineness of the pulp regulated
accordingly.

A disc refiner has other advantages over a Jordan because it is much cheaper
to construct and the discs can be changed quickly not only if they are worn but
also to have the refiner plate shape best suited for the particular material passing
through it and the character of the paper to be made. This gives a mill greater
flexibility in pulp preparation. A disc refiner works on a larger percentage of
the individual fibres than does a Jordan and gives the fibres more fibrillation
rather than cutting.[14] Therefore a stock like waste paper in which the fibres

193

are short already will not be further degraded by passing through a disc refiner.

The Evolution of the Fourdrinier Machine

The speed of a modern paper machine has risen from 700 ft./min. (213 m./min.) around 1900 to 960 ft./min. (300 m./min.) for an ordinary machine and 4,800 ft./min. (1,500 m./min.) on newsprint machines to even 6,000 ft./min. (1,830 m./min.) on some tissue machines today.[15] This has been achieved through better design as well as new or improved materials for building the machines and more accurate manufacture. It was hoped that the introduction of electric motors would enable speeds to be raised but there was so much drag in the old-fashioned sleeve bearings that this proved impossible until anti-friction bearings were fitted to a tissue machine in 1922. The power to drive this reconstructed machine was reduced to a fraction of what was needed before.[16] Very soon other machines were built using anti-friction bearings and older ones converted so that operating speeds more than doubled in the inter-war years. As well as tissue machines, newsprint machines were designed for these faster speeds.

The first effect of this was that speeds were too great to allow the pulp or stuff to flow out of the breast box by gravity but it had to be forced out by pressure. All air bubbles had to be removed before the stuff reached the breast box otherwise they could explode as they were ejected and form holes in the paper on the machine wire. Also the flow of the stuff through the breast box had to be carefully directed so it came out of the discharge slot evenly and did not form eddies. Today vanes may be fitted inside the box to guide the stuff. Because the stuff is forced out under pressure, the size of the discharge slot has to be regulated precisely otherwise the finished paper will be uneven. Instead of a few plates held in position by wing-nuts, today the slice consists of a whole row of valves, each one having its own regulating apparatus which can be controlled by a computer after an initial setting manually.

Speed of removal of water is vital because something like twenty tons of water are used to make one ton of paper although not all of this passes through the wire. The stock flowing out of the breast box will have a consistency of about one per cent of fibres, or one part of fibre to 99 parts of water by weight. The bulk of the water is taken out through the wire by the foils, vacuum boxes and vacuum couch. After the couch, the moisture content is 80 per cent, that is 95 parts of water have been taken away. After the presses, the moisture content is about 60 per cent, that is a further 2½ parts of water have been taken out, almost all of it going through or away with the wet felts. This shows the importance of conditioning and drying these felts. The remaining one part of water is evaporated on the dryers and the dryer felts in the form of vapour but of course this vapour occupies almost 2,000 times the volume

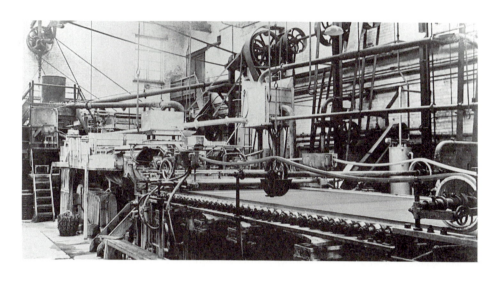

47 & 48. These pictures show nearly sixty years of development on Oakenholt No.3 machine. The top picture was taken in 1928 and the bottom in 1987. In the intervening period, all the pulp delivery system has been changed, the rollers beneath the wire replaced by foils and most of the belt drive changed for electric motors.

195

of the liquid water.[17] At each successive stage, it takes more equipment to remove water and it also costs more.

Although the principle of the Fourdrinier machine remains the same today as it always has, modern machines have been transformed by plastic and man-made fibre components. No longer is the wire supported on rollers but it passes over tapered plastic strips called foils. The narrow point of the taper scrapes the water off the bottom surface of the wire and then, as the wire moves on, more water is sucked through in an area of lower pressure between the nip of the foil and the wire. More than twice as many foil blades can be installed in the space required by the table rolls they replace and, since foils increase drainage capacity, more water can be added to improve stock formation and to increase machine speeds.[18]

The wire itself is no longer copper or bronze but is some form of plastic. Plastic does not corrode or fatigue and can be woven into an open fabric with depth that helps to draw the water through it in the same way as the supporting grid underneath the cover of a hand mould. Plastic wires are much lighter and so can be handled more easily and do not wear as quickly. At the Grove Paper Mill in Disley, metal wires lasted three to four weeks but plastic ones for about a year. Today metal wires are used only on cylinder mould machines and on dandy rolls. Shadow watermarks have to be formed in bronze wire too. These wires, whether metal or plastic, must be woven in a single width and so the looms must be as wide as the paper machines which can be over 300 in. (7.62 m.) today. With as many as 600 individual warp threads in an 8 in. (200 mm.) space, the difficulty of the tasks of setting up and weaving well can be imagined. Today some wires are woven as a continuous tube with the weft running the length of the machine so that for a wire 117 ft. (36 m.) long, the loom must be 58.5 ft. (18 m.) wide.

The problem with the Fourdrinier machine is that the two sides of the sheet of paper will be formed with different characteristics. To begin with, as the stuff pours onto the wire, the short fibres will pass through the holes, leaving the longer ones behind. These longer ones will form a mat which will catch the shorter ones so the next layer consists of a mixture of both. At the top, that surface will consist of a greater proportion of short fibres because the heavier longer ones will tend to settle first. This difference in fibre length has caused trouble on some modern high speed printing machines with the short fibres rubbing off. Then the two sides of the sheet will have different surfaces because one has been formed against the wire and the other not. Pressure against the wet felt or the smooth surface of the wet press rollers will impart differences to the surfaces again. Water removal on the conventional Fourdrinier machine takes place through only the lower surface of the sheet and the wire. If the water could be removed through both sides, not only would it take less time but the paper would be more equal on both sides. While people have struggled to invent ideas to remove these defects over many years, it has been in the last thirty particularly that new types of machine have been introduced.

One way of making the sheet of paper equal on both sides has been to use a pair of Fourdrinier wires so that the sheet is in effect a board. This idea seems to have been patented originally by William Ibbotson in 1875.[19] He arranged the wet ends of two Fourdrinier machines so that the webs from both could be pressed together either at the couch or at the drying cylinders. This idea has been taken up more recently by Tullis Russel in Fife who make high quality printing papers. Their first twin wire or double machine was installed in 1923 and their latest and

49. Top former installation on No. 6 paper machine at Reed Paper & Board's Lower Darwen paper mill.

largest machine, which started operating in 1979, has a capacity of 20,000 tonnes per annum.[20] Such machines can make paper with different colours on either side, such as the backing paper for photographic roll films which are black on one side and often yellow on the other.

That such machines are called twin-wire machines is in some ways a misnomer because now machines have been developed in which a second wire runs on top of the first and sandwiches the pulp between the two. It is possible to put foils or suction boxes on top of the higher wire to remove water upwards. The first machines of this type were the 'Inverform' board machines which will be described in that section. The research team at St. Anne's Board Mill, Bristol, realised that the same principle could be applied to make a single ply paper and so they joined forces with Walmsley's, the machinery manufacturers in Bury. Design started in 1956–8 and the 'Twinverform' machine appeared in 1962–3. Other types of twin wire formers have evolved out of this machine so that today all the paper machine manufacturers offer some type of machine on this principle. On some of the latest high speed newsprint machines, this second wire is brought down to the first part way along the horizontal section of the wet end so there is partial drainage in the traditional Fourdrinier style. Then both wires descend together through curving guides so that the water carries straight on and is in effect centrifuged through the upper wire.

One recent layout of a machine with two wires is the 'vertiforma' developed in America by Black Clawson in about 1966. Here the pulp is directed straight down from the slice into the gap between a pair of rollers round which two sets of wires are passing. As the water is taken off by foils and vacuum boxes on both sides, the wires are brought closer together until they turn round a common roller at the bottom of the machine and then are taken horizontally over more vacuum boxes before separating when the paper sheet can be removed. Examples have been tried in this country, one being installed at Bowater's Mersey mill in 1970 for the production of paper for printing in off-set litho presses where the surface on both sides must be the same. In other layouts, the wires move upwards and in fact there is a whole variety of forming devices available today and no new machine will have a single wire in the traditional Fourdrinier layout.

Generally today the final couch roller is perforated and vacuum applied inside it. The suction couch was patented by Millspaugh Ltd. in 1908 but some years were to elapse before it came into general use. One advantage is that the wire is drawn down by the vacuum onto a moving part of the machine which causes less wear than on vacuum boxes. Others are that higher machine speeds are attainable, guiding the wire is simpler and there is no distortion of the web by crushing. Suction was first applied by Millspaugh Ltd. in 1912 to the press rollers but again the general adoption of suction presses did not follow until about 1924 to 1930.[21] It is usual to make the lower roll the one with the suction. These two developments helped to introduce the high speed newsprint machine.

50 & 51. Comparison of Oakenholt's No.3 machine wet press and drying cylinders. The deckle straps have disappeared but a Dougsen dandy with splash hood has been added. The drying cylinders have been covered with a hood to improve drying and to conserve energy.

The Dry End

The main characteristics of any felt are strength, to withstand the stresses on the machine, finish, to give the required surface on the paper and drainage, to allow the water or vapour to pass through. Before 1950, all felts were made almost exclusively from wool but once again plastics have come to the aid of the papermaker. Man-made fibres can be selected to form fabrics with the most suitable characteristics for the operating conditions both in the wet press and on the drying cylinders. Once again, the length of life of these modern fabrics has been extended for wet felts had to be changed every five days but now only every three months. To make woollen felts wide enough for modern paper machines, Kenyons wove the cloth 60 feet (18.2 m.) wide because it then had to be fulled or milled which compacted it to about half its width. After that, the surface had to be raised to form the nap.[22] Today man-made fibre felts can be formed differently through 'needling' techniques which began to be introduced in 1960.[23] Into a loosely woven structure, a web of fibres can be added which can be designed to give the best results for a high speed newsprint or a multi-vat cylinder board machine. Felts on the drying cylinders have an open weave to allow the vapour to pass through more quickly.

New machines may have from fifty to sixty drying cylinders, with some used for drying the felts. The wider the machine, the more difficult it is to remove uniformly the saturated air right across the width of the dryers and so evenly dry the paper. By the late 1920s, warm air was blown into spaces between the cylinders to absorb the steam.[24] This process is not one of boiling off the moisture but of dissolving it in hot air. The hotter the air, the more moisture it will carry. Hoods are fitted over the cylinders with various types of blowers and extractors in them to assist and accelerate even drying. This air can be re-circulated with the moisture removed to conserve energy. Now high velocity hooded dryers have been developed and systems are available using dryer hoods sectionalised across the width in order to give better control to the drying.[25]

M.G. Machines

Machines on which very light and thin paper is made may have only one drying cylinder. In one way this is reverting to John Dickinson's patent of 1817 but he did not use the surface of the cylinder to give a finish to the paper. In 1843 Lemuel Wright built a cylinder mould machine with one large drying cylinder above the vat. During its passage round the cylinder, the paper was pressed against it by three or more press rolls which also helped to smooth the paper. The paper will adhere to the surface and, as it dries, it will contract so that, if the surface of the cylinder is highly polished, that side of the paper in contact with it will be highly polished too. Wright claimed that this smooth paper surface was far better for writing upon than

52 & 53. The earlier calender stacks and reel at Oakenholt mill have all been removed and instead there are a calender with rolls pressed together with hydraulic cylinders, the computer measuring head and a new reel.

54. The dry end of T. Tait & Sons paper machine No. 4 which started production in June
1986. At 6.84 meters wide trim, it is the widest machine producing fine paper in Britain.

any other previously made and that, for printing purposes, it was excellent too.[26]
The cylinder could be up to 15 ft. (4.5 m.) in diameter although recently much larger
ones have been cast. The surface must not have any blemishes so that the web can
be drawn off it freely. If the paper is too wet, it will not stick onto the cylinder and
if the fibres are too short or too 'free', it may stick to the cylinder in patches and give
a pock-marked finish. Most cylinders are now enclosed and hot air blown in to help
remove the water vapour.[27]

Once again British papermakers do not seem to have installed machines of this
type until the 1880s after they had become popular in America. They could be
linked to either a Fourdrinier wet end or a cylinder mould machine. In this country
they became known as 'Yankee' or 'M.G.' machines because the smoothness of the
cylinder imparts a high machine glaze to one side of the paper. Tissues, paper for
bags which are printed on only one side, thin wrapping paper, some kraft papers
are among the products they produce. Springfield Mill near Bolton installed one in
1924 which was designed for kraft papers at a rate of 40 to 50 tons a week at a speed
of up to 300 ft./min. (91.4 m./min.).[28] A 120 in. (3.4 m.) machine was built by
Walmsley's of Bury for the Disley Paper Mill Co. Ltd. in 1929 and was so successful
that another followed within three years.[29] Both, much modified and improved,

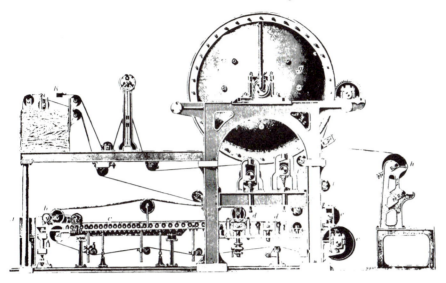

55. An M.G. machine with the single drying cylinder above the Fourdrinier type forming part, c. 1900.

are still running today. These are a far cry from one erected by Reed and Smith at their Wansborough Mill in 1977 with a capacity of 20,000 tonnes per annum in the 36 to 120 gsm. range for papers for stationary, wallpaper and packaging as well as tissues from waste paper pulp.[30]

Board Machines

Improvements have been made at the other end of the weight range, in board production. Speeds on the traditional type of cylinder mould machine with multi-vats for making board remained relatively slow with a maximum speed of 350 to 400 ft./min. (100 to 122 m./min.) compared with 2,300 ft./min. (700 m./min.) for a Fourdinier. In the Dickinson machines, the cylinder in the vat turns against the flow of stock in what is called counter flow. Speeds could be raised by having the cylinder revolving in the same direction as the pulp entered in a uniflow layout. More recently Bramah's style of dry mould machine has been tried with pressure head boxes injecting pulp onto the outside of a rotating cylinder without a vat. These were developed by the Boxboard Research and Development Association in conjunction with St. Anne's Board Mill, Bristol, and the first ones were produced in 1974–5. Vacuum boxes are installed inside such cylinders in the style of Dickinson's early machines and now there is a wide variety of machines built on these principles with over three hundred built on the St. Anne's former type alone.[31]

56. No.5 Multi-ply board machine of C. Davidson & Sons, Aberdeen. The 6-ply wet end comprises a Fourdrinier, Arcu Forma and four Inverform stations capable of producing boards in the caliper range 300 to 750 Microns. The 2.85 metre wide machine manufactures over 100,000 tonnes per annum, making it one of the highest board production units per metre width in Europe.

57. Computer console at Kimberly-Clark's No.4 machine at Prudhoe mill in Northumberland showing the size of modern computer installations.

Pressure of demand for board led in 1958 to the development of a new type of
machine by R.J. Thomas and his research team at St. Anne's Board Mill. The new
machine was termed 'Inverform' because it reversed the method of removing water
from the traditional downward flow to suction from above. The board was formed
between two wire webs onto which the stock was deposited from above. Such a
machine would take a much wider range of furnish than a Fourdrinier so that it
expanded the use of waste paper for board making.[32] The new process was soon
licensed to the Thames Board Mills who built a machine with a capacity of 80,000
tonnes per annum at Purfleet which started production in May 1963. At 1,600 ft.
(500 m.), it was the longest board mill in Europe.[33] It had five inverform units so
it could produce board with up to five different layers which were bonded together

before passing to the drying cylinders. The three inside layers might be made from waste paper pulp with the two outer ones from a wood pulp to give a good printing surface. Other machines have followed elsewhere in both England and Scotland and it is largely due to the adaptability and speed of inverform machines that Britain retained a large share in the greatly expanded market for packaging board in the packaging revolution since the Second World War.

Computer Control

In about 1966, an early English Electric analogue computer was installed at Wolvercote Mill and a second at Grove Mill, Disley. Today there is virtually no mill which does not have a computer control. Measuring heads continually monitor the sheet of paper just before it is reeled up to record thickness and moisture content. Other details recorded include water and steam consumption, steam pressures, machine speed, head box levels and pressure, stocks of raw materials and chemicals, and of course the amount of paper actually made. The computer can detect any changes in the paper much more quickly than former methods which involved testing samples in laboratories. The computer can be linked to the mechanisms controlling the slice openings to let more or less stock onto the wire, the speed of the machine or the temperature in the drying cylinders. All these can be regulated much more closely to produce paper of more even consistency. Also all stocks of all materials can be known instantly. To change the grade of paper, the operator feeds the relevant information into the computer which resets the machine. Now the range of computer control is being extended from the machine to the whole operation of the mill. This will enable orders for paper to be programmed more efficiently, raw materials ordered in advance and the different grades of paper run on the machine at the optimum settings.

The Most Recent Trends

Indigenous Resources

A final feature in the development of the modern paper mill during the 1960s has been the growing importance of the integrated mill where the tree is delivered at one end and appears as paper at the other. When the British coal mining industry introduced hydraulically operated pit props to replace wooden ones, the Forestry Commission no longer had an outlet for the sale of some of its trees, particularly the thinnings. The paper industry has taken advantage of this and new mills have been and still are being constructed to process indigenous supplies of timber. Sitka spruce has a faster rate of growth in this country than in Scandinavia. With the decline in agriculture, more land could be laid down with forests

206

and the paper industry might expand once more using a renewable domestic resource.

In 1963, a special Act of Parliament was passed to provide £8 million to help provide employment in a remote area in a £15 million project by Wiggins Teape to establish a chemical pulp and paper mill at Corpach near Fort William in Scotland. It was to use wood from the forests planted in that area for the production of relatively high quality paper for light weight printings and thin card, mainly supplied to the printing industry. Both pulp and paper mills were on stream in 1966 but the pulp mill has proved to be uneconomic and has been closed down, partly because supplies of local timber were both too costly and insufficient. While the pulping side of that venture has not succeeded, the mill at Sudbrook on the Severn Estuary was converted in the latter part of the 1960s to an integrated mill for the production of semi-chemical paper used in the middle of corrugated boards.

Another mill that has taken advantage of development grants in an area of high unemployment as well as the integrated principle is the Shotton Paper Company. A Finnish company opened a new mill in 1985 close to the river Dee to the north of Chester. Here sitka spruce thinnings are converted into thermo-mechanical pulp

58. The first soaking drum for waste paper in Britain has been running since 1986 at the mill of C. Davidson & Sons, Aberdeen. It drastically reduces the work load of the waste paper pulper by breaking up the bales and pre-wetting the paper before it enters the pulper. This drum is 3.2m. diam. and 10.8m long. It is mounted on an hydraulically operated tilting frame and rotates at 6.5 rev/min.

in special tandem disc refiners in five lines which provide 90 per cent of the steam needed in the 49 drying cylinders. About 600 tons of newsprint a day, or 200,000 tons per annum can be produced on the twin wire machine. The sheet width is over 350 in. (8.75 m.) and the finished reel holds up to 26 tonnes of newsprint.[34] Although this mill has only 250 people on its payroll, it has been estimated that up to 1,500 are employed indirectly through transport services and the like. A second machine is to be built on this site. The opening of this and other mills and machines has seen production of newsprint rise from a 'low' of 80,000 tonnes in 1983 to nearly half a million tonnes in 1986, about a third of the consumption in the United Kingdom.

A project which will be completed in 1989 is another integrated mill again being developed by Finns.[35] It is the Caledonian Papermill a little to the south of Glasgow. 500,000 tons of spruce wood from Scotland will be taken each year and turned into light-weight coated paper suitable for glossy magazines, a quality not being produced in this country at present. A chemical process will be used so that the pulp will be high quality which is needed in this sort of paper. The power will be taken from the electricity grid because several other large companies have closed works in the region and the Electricity Board has excess capacity. The site is closer to better transport facilities than Fort William which makes the scheme economic.

Waste Paper

The most important indigenous 'raw' material for the British paper mills today is waste paper. This has long been a source of fibre for board manufacture and we have seen how Matthias Koops tried to use it for making good quality paper around 1800. Gradually better ways of grading the waste, treating it to remove printing inks and colours, and also cleaning it to remove dirt and waste matter, have been evolved so that papers made from waste are moving up-market into the better qualities. Waste paper can vary greatly in its character. Some will be printed, some will contain filler for a smooth glossy surface, some will be wood pulp, some chemical. Leaves may have been bound together with wire staples or sewn with thread, envelopes will have gummed edges and there will be various qualities of sizing and coatings. Dry waste paper must be converted into a wet papermaking pulp, free from bundles of unseparated fibres and free from all impurities. Various systems are employed, depending upon the type of waste, but some of the stages are outlined below.

The bales of waste are dropped into a hydro-pulper where they are broken up in water by a rapidly rotating rotor. While some of the large impurities such as wire, nails and bolts can be removed in the pulper, a preliminary cleaning stage may have to follow before the pulp is passed through some form of disintegrator, probably a type of disc refiner to separate all the fibres. To clean the pulp, it will be passed through a row of centrifugal cleaners, each one consisting of a cylinder on top of a cone. The stock mixture is squirted into the top at a tangent so it spins downwards and flows back up in the centre of the descending spiral. The heavier grit and metal pieces are thrown to the outer walls, slide down and are collected at the base where they can be drawn off. Screens will remove those impurities which have the same density as the pulp such as some plastics, string or synthetic fibres. De-inking stages will follow next which may be done by washing or floatation. For washing, chemicals are added during the pulping to assist in the release of the printing ink from the fibres. The ink and any filler such as clay is washed through a screen, leaving behind the fibres which are removed from the screen and sent to the paper machine.

59. Aerial view of Prudhoe Mill, Northumberland, showing the extent of a modern mill.

Floatation systems use different chemicals so the ink is collected in a frothy scum on the surface of a tank while the cellulose fibres remain in suspension and are passed on for possible further washing and bleaching. After this, the fibres will be made into paper in the usual way.

During the First World War, waste paper supplemented the supply of other pulp and much was collected from consumer industries and the public. From this time onwards, waste paper became a major raw material for cardboard and the Thames Board Mills at Purfleet by 1939 had 200 local authorities in England acting as collectors. There were similar link-ups in Scotland with Davidson and Radcliffe so that in the 1930s, about 800,000 tons per annum of waste paper were salvaged of which 600,000 tons were used at home and 200,000 exported.[36] The Second World War, like the First, saw drives to save waste paper and subsequently waste paper collection has been organised on a large scale. The consumption of waste paper as a percentage of paper and board production was 38.5 per cent in 1968 which had increased to 54 per cent in 1986.[37] Waste paper is used in more than half of the mills in this country.

CONSUMPTION OF WASTE PAPER

Year	'000 tonnes
1971	1,800
1972	1,900
1973	2,000
1974	2,100
1975	1,700
1976	2,014.2
1977	2,084.2
1978	2,092.1
1979	2,184.1
1980	2,008.5
1981	1,932.0
1982	1,853.8
1983	1,827.0
1984	2,002.7
1985	2,067.1
1986	2,147.4[38]

The problem with waste paper has been to secure an orderly market. The upturns and down turns in papermaking reflecting the general economy have hit waste paper collection particularly hard. In a slump, the price which people can obtain for their waste has made it not worth the trouble of collecting and it is difficult to persuade them to save paper again when times improve. This has applied equally to the

60. The new coating machine for carbonless copying paper, part of a recent £15 million investment by Wiggins Teape at Ely mill, Cardiff.

This has applied equally to the collections organised by the local authorities.

It has been estimated that it takes an average of 17 trees to produce just one tonne of paper, on which 7,000 copies of a daily newspaper can be printed. One issue of the Daily Mirror would take about 10,000 trees if it contained no fibres from re-used waste paper. In 1974, Britain would have consumed a forest about the size of Wales to meet its paper requirements[39] so waste paper can make a substantial contribution to conservation of scarce resources. In 1974, Britain was the second highest consumer of waste paper in Europe after West Germany and was fourth in the world, behind the United States, Japan and West Germany. While virgin fibre may still have to be added to make certain grades of paper, the British paper industry would be much smaller without the resource of waste paper. Waste paper gives domestic mills a better chance of competing with foreign producers with larger forest resources as well as putting to good use a waste product which would otherwise require costly disposal. It also supports a growing sector of the packaging and board industry for which there is no economic raw material substitute.

Tissue Paper

The soft tissue industry started in this country in the 1920s. One of the first companies in this field was Kimberly-Clark of Wisconsin, United States of America, who had developed cellulose wadding as a substitute for cotton wool during the First World War. It was more absorbent, soft but strong, and soon American nurses were making sanitary towels from it. Kleenex tissues were introduced in 1926 and were being sold in Britain that year. During the inter-war years, there was a small tissue industry in this country using M.G. machines. It was not until after the Second World War that the spectacular growth began. Between 1946 and 1981, production capacity was increased from 13,000 to over 400,000 tonnes, averaging 10 per cent compound annual growth for most of the period. The tonnage is now over half a million.[40]

Of the new tissue machines installed since 1970, five are 16.4 ft. (5 m.) wide and, with speeds of well over 4,000 ft./min. (1,200 m./min.), each has a capacity of more than 30,000 tonnes a year. Recent investment by Kimberly-Clark at the Prudhoe Mill in Northumberland has resulted in a capacity of 80,000 tonnes a year, requiring 60,000 tonnes of high grade waste and 30,000 tonnes of wood pulp. The main investment here has been in the waste processing plant to up-grade lower types of waste for the production of toilet tissues and industrial wipes. About a quarter of the investment has been spent on an effluent disposal plant. In 1986, the total sales of toilet tissue was valued at £396 million, of facial tissues £116 million, and of kitchen towels, £84.5 million.[41] Traditionally about 90 per cent of tissue consumption is manufactured in this country.

Coated Papers

Another section of the paper industry which is growing in importance is coated papers. The reasons for sizing paper have been discussed earlier and some methods explained. While the basic principles are the same for coating paper, techniques have had to be improved to cope with the high speeds demanded today. Coating may be done either as part of the actual papermaking machine or on a separate machine. The advantages for on-line coating are that the sheet will not be completely dried before receiving the coating and does not have to be handled between machines. This is the system to be employed at the new Caledonian Mill.

Essentially coating will give the paper a smoother surface for printing and possibly a different colour. Clay is one substance used for making art papers with a high gloss. Starch, latex or other binders, and optical whitening agents are some other coatings. One coating which has been extremely successful is that for making carbonless copying paper. Because the coating mixture has a high proportion of water, the base paper has to be of the right absorbancy to receive it and also strong enough not to tear while it is wet. Hence the need for a chemical pulp which is stronger than mechanical. The paper may be coated on one or both sides. The sheet of paper passes through the coating substance which is levelled by either an air knife or doctor blade. If it is being done on a papermaking machine, the coating unit will be placed at the end of the first set of drying cylinders and then, after being coated, the paper will pass round more drying cylinders. A stack of calender rolls will be placed at the end before the reel so that, if required, the paper can be given a high gloss. High speed calendering may be done separately. The extent of the market for light weight coated paper may be seen in the ever growing numbers of colour magazines, such as the Sunday colour supplements, holiday brochures, mail order catalogues and the like.

Paper Today

Today it would be hard to imagine a world without paper. From the day we are born to the day we die we use it continually. Surgical instruments for delivering babies at the hospital will have been wrapped in sterilized paper to prevent contamination. Disposable nappies, paper towels, incontinence pads, all are made from tissues. At home, we have wallpaper, paper handkerchiefs and napkins, toilet rolls, paper plates and cups, milk cartons, decorative laminates such as formica, books, newspapers, magazines and writing materials. Food comes wrapped in greaseproof paper or other forms of paper packaging as well as in paper bags. The coffee we drink may have been filtered through paper. Egg boxes, cartons and many sorts of storage containers are made from paper and board. Children's games, such as snakes and ladders, are mounted on boards and of course there are playing cards, paper hats, fireworks and sleeves for records. We cannot go on holiday without money in the form of travellers cheques or bank notes. We cannot even start our journey without the tickets for the train or aeroplane. We will not be able to go very far without guidebooks and maps. The computer at work, accounts, bills, postage stamps, archives and records, none of these would be possible without paper. Industry has a whole range of uses for paper in filtration, insulation for both heat and electrical systems, damp proofing, cleaning materials, etc. The list could be extended many times. Five hundred years ago John Tate had no inkling of the amazing industry he was about to launch.

Bibliography

Anon, *East Lancashire Paper Mill Ltd., One Hundred Years of Progress, 1860–1960*, Privately printed 1960.

Balston, T., *William Balston, Paper Maker, 1759–1849*, Methuen, London, 1954.

Balston, T., *James Whatman, Father and Son*, Methuen, London, 1957.

Barrett, T., *Japanese Papermaking, Traditions, Tools and Techniques*, Weatherhill, New York, 1983.

Blagden, C., *The Stationers' Company, A History, 1403–1959*, Allen & Unwin, London, 1960.

Bolam, F., Ed., *Paper Making, A General Account of Its History, Processes and Applications*, Technical Section, B.P. & B.I.F., London, 1965.

Carter, H., *Wolvercote Mill, A Study in Paper-Making at Oxford*, Clarendon Press, Oxford, 1957.

Chater, M., *Family Business, A History of Grosvenor Chater, 1690–1977*, Privately printed 1977.

Churchill, W.A., *Watermarks in Paper in Holland, England, France, etc., in the Seventeenth and Eighteenth Century and Their Interconnections*, Paper Publications Society, Amsterdam, 1935.

Clapperton, R.H., *The Paper-making Machine*, Pergammon Press, Oxford, 1967.

Clapperton, R.H. & Henderson, W., *Modern Paper-making*, Blackwell, Oxford, 1958.

Coleman, D.C., *The British Paper Industry, 1495–1860*, Clarendon Press, Oxford, 1958.

Evans, J., *The Endless Web, John Dickinson & Co. Ltd., 1804–1954*, Cape, London, 1955.

Grant, J. & Young, J.H., *Paper and Board Manufacture, A General Account of Its History, Process and Applications*, Technical Section, B.P. & B.I.F., London, 1978.

Green, T., *Yates Duxbury & Sons, Papermakers of Bury*, Privately printed 1963.

Heawood, E., *Watermarks Mainly of the Seventeenth and Eighteenth Centuries*, Paper Publications Society, Hilversum, 1950.

Hunter, D., *Papermaking, The History and Technique of an Ancient Craft*, Dover, New York, 1978 edn.

Ketelby, C.D.M., *Tullis Russell, 1809–1959*, Privately printed 1967.

Krill, J., *English Artists Paper, Renaissance to Regency*, Trefoil, London, 1987.

Labarre, E.J., *Dictionary and Encyclopaedia of Paper and Paper-making*, Swets & Zeitlinger, Amsterdam, 2 edn. 1952.

Loeber, E.G., *Paper Mould and Mouldmaker*, Paper Publications Society, Amsterdam, 1982.

Lyddon, D., & Marshall, P., *Paper in Bolton, A Papermakers Tale*, Privately printed 1975.

Muir, A., *The British Paper and Board Maker's Association, 1872–1972*, London, 1972.

Oman, C.C. & Hamilton, J., *Wallpapers, A History and Illustrated Catalogue of the Collection in the Victoria and Albert Museum*, Sotheby, London, 1982.

Mackenzie, A.D., *The Bank of England Note, A History of Its Printing*, Cambridge University Press, 1953.

Mandl, G.T., *Three Hundred Years in Paper*, Privately printed 1985.

Reader, W.J., *Bowater, A History*, Cambridge University Press, 1981.

Shears, W.S., *William Nash of St. Paul's Cray, Papermakers*, Privately printed 1950, revised 1967.

Shorter, A.M., *Paper Making in the British Isles, an Historical and Geographical Study*, David & Charles, Newton Abbot, 1971.

Shorter, A.M., *Paper Mills and Paper Makers in England, 1495–1800*, Paper Publications Society, Hilversum, 1957.

Sugden, A.V. & Edmondson, J.L., *A History of English Wallpaper, 1509–1914*, Batsford, London, 1925.

Thomson, A.G., *The Paper Industry in Scotland, 1590–1861*, Scottish Academic Press, Edinburgh, 1974.

Tillmans, M., *Bridge Hall Mills, Three Centuries of Paper and Cellulose Film Manufacture*, Privately printed, 1978.

Watson, N., *The Last Mill on the Esk, 150 Years of Papermaking*, Scottish Academic Press, Edinburgh, 1987

Glossary

Agitator – a revolving paddle used in stuff chests or vats to mix and keep the pulp stirred.

Air-dried – is a term applied to machine-made paper which is passed over skeleton drums and dried by air, usually hot, circulated through them.

Air-knife coating – the so-called air-knife acts on the principle of a doctor blade and uses a thin, flat jet of air to remove the excess coating from a wet sheet which has just passed through the coating zone.

Alum – is a term used by papermakers for aluminium sulphate for sizing. Correctly, it is a general name of a group of double sulphates, all of which crystallize in the same form such as potash alum.

Aluminium sulphate – is used in the sizing of paper. It is added to animal size or gelatine to stabilize the consistency, to act as a preservative by arresting the formation of destructive bacteria and to help render the gelatine impervious to ink.

Animal gelatine or size – is produced by boiling the waste pieces of hides, hooves, bones etc. in copper-lined steam-jacketed heaters. The gelatine helps to prevent ink and water penetration into the paper.

Antiquarian – was a size of drawing paper introduced by the younger Whatman in 1773. It was the largest size sheet of paper made by hand in the west and was standardized at 53 x 30 in.

Apron – is a sheet of oiled cloth, leather or rubber which bridges the gap between the breast box and the moving wire on a Fourdrinier paper machine so that the pulp is delivered evenly onto the wire.

Beater or Hollander Beating Engine – is the device generally assumed to have been invented in the middle of the seventeenth century in Holland and which superseded the older method of preparing the pulp by hammering or stamping the rags in a mortar.

Beater roll – is a cylinder or drum with knives set around its circumference which cut up the rags and fibrillate the fibres against the bedplate set in the bottom of the beater trough.

Bedplate – in a stamper it is a flat plate of iron in the bottom of the trough against

217

which the rags are pounded. In a Hollander beater it has ridges or knives set across it and is placed in the bottom of the Hollander beater so that the pulp has to pass between it and the roll above it. The distance between the two determines the fibre length of the pulp.

Board – is a thick sheet of paper. It may be homogeneous througout or may be made from layers of paper, either pressed together while wet or glued to form the sheet.

Board machine – is similar to the cylinder mould machine but has from two to seven cylinder moulds in a line to form multi-ply boards.

Breaker or Breaker engine – gives the primary reduction of the pieces of rags to make them smaller and fit for their final beating in the Hollander.

Breast-box – is the part of the papermachine from which the pulp issues onto the moving wire. On most recent machines, the pulp is forced out of the breast-box under pressure.

Breast roll – is the roller around which the wire on a Fourdrinier machine passes under the breast-box just before the stuff is poured onto it.

Calender – is a set or stack of rollers or rolls between which the paper passes and is smoothed by their weight. The calender is placed at the end of the papermachine while the super-calender is separate. Both may have heated rolls.

Cardboard – is a term applied to thick, stiff papers, or stiff board produced by pasting together a number of layers of paper.

Cellulose – is the basic substance of paper manufacture, the chemical formula being $C_6 H_{10} O_5$. It is the predominant constituent of plant tissues from which it must be separated before it can be used.

Chain lines – are the more widely spaced watermark lines across the narrow way of the sheet. They are caused by the tying wires which bind the laid lines into the cover of the mould.

Chemical wood pulp – is wood reduced to pulp by a chemical process, e.g. by boiling or digesting with either caustic soda, caustic soda and sulphate or soda or with bi-sulphite of lime.

Coating – is the term applied to mineral substances such as china clay which are used to cover the surface of the paper to make it more suitable for some methods of printing.

Couch – is the action of transferring the sheets of newly formed paper from the hand mould onto a felt blanket so that the water can be pressed out. The Coucher is the man who carries out this operation.

Couch-rolls – on a Fourdrinier papermachine are situated at the end of the moving wire from which the paper is transferred onto a felt blanket.

Cover – is the wire surface of a hand mould through which the water drains, leaving the fibres behind to form a sheet of paper. It is also applied to the surface of a dandy roll.

Cylinder – is a term indiscriminately applied to various kinds of rolls or drums on papermachines. More particularly the term is applied to the steam heated cylinders used for drying the web of paper.

Cylinder machine, cylinder mould machine – was invented by John Dickinson in 1809 and has a cylinder covered with wire through which the water drains, leaving the pulp on the surface. The cylinder is partially immersed in a vat of pulp. It has been developed into board machines and machines for making paper with complex watermarks.

Dandy roll – is a light skeleton roll or cylinder covered with wire which presses gently on the paper while still wet. It helps to improve the formation of the sheet and can be used to impress a watermark on paper made on a Fourdrinier paper machine.

Deckle – in hand papermaking is the removable frame around the mould which helps to retain the pulp on the mould's surface while the water drains through. On a Fourdrinier machine, the deckle straps perform the same function on the moving wire.

Deckle edge – when the deckle is removed from the mould and the paper is couched, the edge of the paper becomes thinned out in a slightly wavy line which is the true deckle edge, an effect found only in handmade paper.

Devil – is a machine for removing dust and dirt from rags or esparto grass (also called a 'willow').

Digester – is the vessel in which wood chips, esparto grass or rags are boiled with chemicals. It can be stationary or revolving, horizontal or upright, cylindrical or spherical according to the system used.

Doctor – is a thin metal blade which scrapes excess liquid or fibres off a roller to help maintain a smooth surface.

Dry cylinder machine – is one where the pulp is poured onto the surface of the cylinder so that the water drains away through the cover on the cylinder.

Dry end – is the term for the drying section of the papermachine consisting of the drying cylinders, calender, reel, etc.

Duster – a mechanical contrivance, usually consisting of a revolving drum of wire mesh for opening out rags, esparto, etc. and getting rid of the dirt.

Engine – originally a term applied to any machine but in a papermill particularly to the Hollander beater. Hence *Engine-sized* refers to the addition of the sizing materials during the beating stage.

219

Felts – used in papermaking are a woven material of either cotton or wool with a raised surface which supports the wet sheet of paper during the stages of removal of water and on a machine the subsequent drying.

Flong – a board used for forming stereotype moulds.

Foils – tapered strips of plastic fitted under the moving wire of a Fourdrinier machine to scrape off the water and drain the sheet more quickly.

Fourdrinier – is the name applied to the normal type of papermachine after the brothers who financed its early development.

Furnish – is the material from which paper is manufactured.

Gelatine – a nitrogeous constituent of skin, bones and hooves of animals, used as a size.

Glaze – is the gloss or polish on a sheet of paper.

Grinder – the machine used to prepare mechanical wood pulp, consisting of a revolving grindstone against which the logs are pressed to disintegrate them.

Ground wood – is pulp produced by grinding wood.

Half stuff – any partially broken or beaten source of fibres for papermaking is termed half stuff.

Hollander – see Beater.

Intermittent board machine – produces thick sheets of board by winding the paper as it comes from either a Fourdrinier or a cylinder machine onto a roller in layers. When sufficiently thick, the web is cut along the length of the roller and pulled off.

Jordan – a machine for reducing or making finer the stock or pulp before it passes to the papermachine. It has a cone with knives around its circumference which rotates within another, also fitted with knives. It was invented by J. Jordan, Hartford, Connecticut, U.S.A. in 1859.

Knotter – an appliance with vibrating screens for removing knots or lumps from the pulp.

Kollergang – the German name for an edgerunner used for pulping materials for papermaking.

Laid lines – are the close light lines in laid paper formed by the laid lines of the hand mould or dandy roll.

Laid papers – are those which, when held up to the light, have a ribbed or lined effect due to the paper being made on a mould made with a cover of closely placed parallel laid wires to form the sieve through which the water drains.

Layer – is the person who separates the sheets of hand made paper from the felts on which they have been pressed.

Liquor – a general term for chemical solutions, but in the paper industry chiefly used for the alkaline solutions.

Look-through – structural appearance of a sheet of paper observed when viewed by transmitted light.

Mechanical wood pulp – is the pulp prepared by purely mechanical means, e.g. by grinding the logs of wood.

M.G. or Yankee machine – has a single highly polished steam-heated drying cylinder to which the sheet of paper adheres as it dries and receives a smooth surface on the side in contact with the cylinder.

Mould – a device consisting of a rectangular wooden frame across which is stretched a covering of wire to act as a sieve or strainer. The sheet of paper is formed on the surface by dipping the mould in a vat of fibres suspended in water which drains away through the cover when the mould is lifted out.

Newsprint – the name applied to paper for printing newspapers. It is the cheapest type made.

Pack or wad – may be either the pile of wet sheets assembled by the layer which he has separated from the felts or a small number of sheets piled up ready for glazing.

Paper-hanging – an obsolescent term for wallpaper.

Pasteboard – a general term for cardboard formed by pasting fine papers to either side of middles of inferior quality.

Plate glazing – a method of producing a smooth surface on sheets of generally handmade paper by placing them between polished plates of zinc or copper and passing them back and forth with slight friction between pressing rollers.

Post – a term applied to a pile of sheets, normally 144 but varying in number, of wet pulp, fresh from the mould, just made into paper couched with alternate felts and ready for pressing.

Potcher – one of the series of beaters or engines used in washing and preparing especially esparto into a pulp.

Presse pâté – is a machine practically identical to the wet end of a Fourdrinier papermachine used to turn wood pulp into sheets which can be transported to another mill for making into the final paper.

221

Pulp – the aqueous stuff containing disintegrated fibrous material from which paper is made.

Rags – the original material from which paper was made but now rarely used except for papers of the highest quality. Only certain types of fibres could be used, e.g. linen, cotton, jute and some types of hemp ropes.

Ream – the term used for a quantity of sheets of paper, at one time 480 but this could vary depending upon the type of paper and today is often 500. It probably bore some relation to the number of sheets a vatman could make in a day.

Reel – is the general term for the revolving frame or drum which receives the paper coming off the machine.

Refining – was originally beating out any lumps left in the pulp before it passed to the papermachine but is now used for the final beating of wood pulp. Hence the *Disc refiner* is the machine which today beats most of the pulp for papermachines with rotating ribbed discs or plates between which the fibres pass.

Retting – a term applied to soaking flax in water in order to rot the hard stems which could then be broken to leave the fibres. In papermaking the rags were soaked with water so they rotted and became easier to beat.

Ribs – are the thin bars of wood which support the wire cover of a hand mould. They normally run across the narrow way and in a laid mould support the chain lines.

Shadow zone – is a thicker area in a sheet of paper formed either side of the supporting ribs of a hand mould because the ribs have drawn the water out of the pulp through the single layered cover, so attracting more fibres towards them.

Shake – is the term applied to a sideways movement of a hand mould or the wire of a papermachine to interlock the fibres while they are still suspended in the pulp.

Size – originally a solution of glue or gelatine but later any substance that reduces the rate at which paper treated with it absorbs water.

Slice – is the opening through which pulp is poured onto the wire of a Fourdrinier machine.

Soda process – is the alkaline treatment of wood for the production of chemical wood pulp by digesting the fibres under pressure with a solution of caustic soda.

Stamper – the early form of device with hammers falling into a mortar used for pulping the materials for making paper.

Stationer – originally a tradesman who had a station or shop, as distinct from an itinerant vendor, formerly a book-seller, or publisher, or both, but now only a tradesman who sells writing materials, chiefly paper.

Strainer – a device consisting of screens to keep back impurities from the pulp passing to the paper.

Strawboard – a cheap coarse board made on a multi-cylinder machine from incompletely cooked straw. Straw paper is a cheap wrapping paper made from unbleached straw pulp.

Stuff, whole stuff – is paper stock or pulp ready for making into paper.

Stuff chest – is the large supply chest or tank provided with an agitator in which the stuff is stored before passing to the vat or machine.

Suction or vacuum boxes – are placed under the further end of the wire on a Fourdrinier machine to draw out the water from the pulp or paper passing over them.

Suction couch roll – is a perforated revolving bronze shell passing over a suction box which serves to further extract water from the sheet of pulp or paper just before it leaves the wire of a Fourdrinier machine.

Sulphate process – is a method of cooking wood chips generally in sulphate of soda to produce chemical wood pulp. It was first introduce by Dahl in 1884.

Sulphite process – invented by Tilghmann in 1863 / 6 and is the process of pulping wood with sulpurous acid and its acid salts (bisulphite of lime, magnesia or soda) in clossed vessels at high pressure.

Tearing wire – is a thick wire affixed to a hand mould to permit the sheet of paper produced on it to be torn along the position of this wire. It can be applied to a cylinder mould as well.

Thermo-mechanical pulp – is made by heating the chips of wood under pressure as they are passed through the first stage of the refining process.

Three dimensional, embossed or shadow watermarks – are formed on a woven wire cover which has been pressed into higher or lower areas to form a pattern. The lower areas form dark and the higher areas lighter parts of the watermark and the contours give grades in between. Portraits and pictures can be made with this technique which is often used in security papers.

Tub-size – is sizing applied after the sheet of paper has been dried, by soaking the paper in a solution of hot gelatine and alum.

Twin-wire – a name applied to duplex paper made on a pair of Fourdrinier machines but joined while still wet in such a way that the two wire sides come together so that the surfaces on both sides are the same.

Vat – originally the tank containing the stuff in which sheets of hand made papers are made but later applied to the tank in which the cylinder of a mould machine is partially immersed.

Vatman – the man who forms the sheet of paper by dipping the mould into the vat and then lifting it out, forming the sheet of paper on top.

Watermark – is a contrived thickening or thinning in a sheet of paper which give darker or lighter areas in the paper as it is being made. These become visible when the sheet is held up to the light.

Wet end – general term for the portion of the papermachine on which the pulp becomes the sheet of paper.

Wire – is short for machine wire and is that moving part of the Fourdrinier machine on which the sheet of paper is actually formed.

Wire profile – is the pattern made from bent wires which form the watermark. They are sewn onto the surface of the cover of a hand mould and protrude into the pulp, causing thinner areas which show as lighter lines in the sheet of paper.

Wire mark – the diamond shaped pattern of the papermachine wire, seen on the wire side or in the look-through of a sheet of paper.

Wove mould – the cover of a mould made from wire woven like a piece of cloth.

Wove papers – are those which do not show any marks of laid or chain lines and so are much smoother than laid papers.

Notes

CHAPTER I   This Paper Thynne

1. D.C. Coleman, *The British Paper Industry, 1495–1860*, 1958, p. 7, note 2.
2. ibid, p. 10.
3. H. Gachet, 'Some Remarks and Reflections Concerning the Introduction of Paper and its Manufacture in the Mediterranean World', typescript, July 1976, p. 7.
4. M.A. Peraudeau & E. Maget, *Le Moulin à Papier Richard de Bas*, 1973, p. 5.
5. Gachet, 'Remarks', p. 6.
6. Gachet, 'Remarks', p. 5 and O. Valls, *The History of Paper in Spain, X–XIV Centuries*, Vol I, 1978, p. 102 f.
7. Wang Juhua and Li Yuhua, 'New Findings Confirm Paper's Origins', *I.P.H. Year Book*, Vol 6, 1986, p. 191 f. and Wang Ren Hong and Yuan K. Yung, 'Talking About Ancient Chinese Paper History', typescript.
8. E. Childs, *William Caxton, A Portrait in a Background*, 1976, p. 40.
9. L. Lyell & F.D. Watney, *Acts of Court of the Mercers' Company: 1453–1527*, 1936, pp. xiv–xvi & 631.
10. W.A. Churchill, *Watermarks in Paper in Holland, England, France, etc. in the XVII and XVIII Centuries and their Interconnection*, 1935, p. 39.
11. E. Heawood, *Watermarks, Mainly of the Seventeenth and Eighteenth Centuries*, 1950, p. 24.
12. D. Hunter, *Papermaking, The History and Technique of an Ancient Craft*, 1978 edn., pp. 270 & 279.
13. A. Stevenson, 'Tudor Roses from John Tate', *Studies in Bibliography*, Vol 20, 1967, p. 15.
14. J.B. Powell, *A Treatise on Papermaking*, 1868, see A.H. Shorter, *Paper Mills and Paper Making in England, 1495–1800*, 1957, p. 18.
15. Shorter, *Paper Mills*, p. 44.
16. Churchill, *Watermarks*, p. 39.
17. See Hunter, *Papermaking*, p. 116, Shorter, *Paper Mills*, p. 174 and Victoria County History, Hertfordshire, Vol IV, p. 256.
18. Stevenson, 'Tudor Roses', p. 17.

19. I am indebted to the staff at the British Library for their help providing books with Tate's paper.
20. Shorter, *Paper Mills*, p. 44.
21. Stevenson, 'Tudor Roses', p. 20.
22. I wish to thank the archivist at Hertford County Record Office for showing me this sheet of paper, ref. D/EAS 4191.
23. J.E. Cussans, *History of Hertford*, 1870–81, Vol II, Sec. 2, p. 52.
24. V.C.H., Herts., Vol IV, p. 256.
25. Hertford Library, typescript.
26. Stevenson, 'Tudor Roses', p. 33.
27. M. Beazeley, 'Tracings of Watermarks at Canterbury Cathedral', *British Museum*, and Stevenson, 'Tudor Roses', pp. 22–3.
28. H.R. de Salis, *A Chronology of Inland Navigation in Great Britain*, 1897, p. 3.
29. J.B. Priestly, *Navigable Rivers, Canals, and Railways, Throughout Great Britain*, 1831, p. 379.
30. V.C.H., Herts., Vol III, p. 472.
31. Hertfordshire County Record Office, D/EX 307 T2, Conveyance 30 Sept. 1867.
32. ibid, D/EX [50]1 & 2.
33. ibid, D/EX [50]1 & 2.
34. V.C.H., Herts., Vol III, p. 493.
35. L Turnor, *History of the Ancient Town and Borough of Hertford*, 1830, p. 62.
36. ibid, frontispiece.
37. W. Branch Johnson, *Industrial Archaeology of Hertfordshire*, 1970, p. 19.
38. Hertford Library, typescript.
39. J. Tate's Will, *Public Record Office* and Johnson, *Hertfordshire*, p. 55 and V.C.H., Herts, Vol IV, p. 256.
40. Tate's Will and Hunter, *Papermaking*, p. 116.
41. Shorter, *Paper Mills*, p. 28, quoting W.S. Gentleman, *A Compendious or briefe Examinacion of certayne ordinary Complaints of divers of our Countrymen in these our days*, 1581.
42. ibid, p. 28, see *A Discourse of the Common Weal of this Realm of England*, c. 1549.

CHAPTER II   Paper of High Grade

1. Y. Kim, 'Papermaking by Hand in Korea', *International Paper Historians Year Book, 1983–4*, Vol 4, p. 254.
2. O. Valls, *The History of Paper in Spain, X–XIV Centuries*, Vol I, 1978, p. 68.
3. J. Evans, *The Endless Webb, John Dickinson & Co. Ltd., 1804–1954*, 1953, p. 129.

4. D. Hunter, *Papermaking, The History and Technique of an Ancient Craft*, 1978 edn., pp. 144–9.

5. L.C. Hunter, *Waterpower in the Century of the Steam Engine*, being Vol I of *A History of Industrial Power in the United States, 1780–1930*, 1979, p. 9.

6. Lt. Col. W. Ironside, 'Uses of the *Sun* and Manufacturing of the Hinostan paper', see Dhrampal, Ed., *Indian Science and Technology in the Eighteenth Century*, 1971, pp. 174–8, taken from *Philosophical Transactions*, Vol 64, 1774, pp. 99–104.

7. J. Trier, *Ancient Paper of Nepal*, 1972, p. 87.

8. Valls, *Paper in Spain*, pp. 181 & 227.

9. see C. Pels and H. Voorn, *De Schoolmeester, World's Last Wind Paper-Mill at Westzaan*, and H. Voorn, 'The Last Wind-Driven Papermill "The School-master" ', *I.P.H. Year Book*, Vol 6, 1986, pp. 202 f.

10. A.F. Gasperinetti, 'Paper and Paper Mills', Zonghi, *Watermarks*, 1960, p. 70.

11. A. Blun, *On the Origin of Paper*, (Trans. H.M. Lyndenberg, New York 1934). pp. 34–5.

12. R.L. Hills, 'Papermaking Stampers: A Study in Technological Diffusion', *International Paper Historians Year Book, 1984*, Vol 5, pp. 67–88.

13. Surviving examples of stampers can be seen in the open air museums at Arnhem, Holland, Hagen, West Germany, at the Richard de Bas Papermill and Museum, Ambert, France and the Paper Museum, Capellades, Spain. A.H. Shorter, *Paper Mills and Paper Makers in England, 1495–1800*, 1957, p. 69 for the reference to Huxham Mill.

14. For a full account of papermaking in Japan, see T. Barrett, *Japanese Papermaking, Traditions, Tools and Techniques*, 1983.

15. E.G. Loeber, *Paper Mould and Mouldmaker*, 1983, pp. 6 & 25.

16. Hunter, *Papermaking*, p. 178.

17. J. Munsell, *Chronology of the Origin and Progress of Paper and Paper-making*, 5th edn. 1876, reprint 1980, p. 69.

CHAPTER III   The Art of Watermarking

1. D. Hunter, *Papermaking: The History and Technique of an Ancient Craft*, 1978 edn., p. 260.

2. *Northumberland Record Office*, ZSW/590, Memo by Sir John Swinburne, 1794.

3. Prof. E. Kirchner Collection, No. 44,285, Deutsches Museum, Munich, Box 1, Sheet 7.

4. W. Weiss, 'Concerning Shadowless Laid Handmade Paper', (Typescript), note 18 quoting from J.G. Krunitz, *General Economic and Technological Encyclopedia*, article on *Paper* p. 654, Berlin 1807.

5. See samples in the F. Wakeman Collection of papers in the National Paper Museum, Manchester and J. Krill, *English Artists Paper, Renaissance to Regency*, 1987, p. 83.
6. Patent 3068, 1807.
7. Loeber, *Paper Mould and Mouldmaker*, 1982, p. 31.
8. ibid, p. 33.
9. Hunter, *Papermaking*, pp. 262 & 474.
10. For sizes of the sheets, see Hunter, *Papermaking*, pp. 136–8.
11. E. Heawood, *Watermarks, mainly of the Seventeenth and Eighteenth Centuries*, 1950, p. 36.
12. Act 34 Geo. III, c. 20, see T. Balston, *James Whatman, Father and Son*, 1957, pp. 99–100.
13. Patent 3598, 1812.
14. ibid.
15. Bank of England Leaflet, *Some Glimpses of the 'Old Lady's' History*.
16. Sir J. Clapham, *A Concise Economic History of Britain from the Earliest Times to A.D. 1750*, 1949, p. 272 and A.D. Mackenzie, *The Bank of England Note, A History of Its Printing*, 1953, pp. 7–8.
17. A.H. Shorter, *Papermaking in the British Isles, An Historical and Geographical Study*, 1971, pp. 51–2.
18. Hunter, *Papermaking*, pp. 282 & 536.
19. Patent 4419, 1819.
20. Hunter, *Papermaking*, pp. 284–94.
21. Patent 4419, 1819.
22. J. Munsell, *Chronology of the Origin and Progress of Paper and Paper-making*, 5th edn., 1876, reprint 1980, p. 152.
23. Mackenzie, *Bank of England*, p. 52.
24. Hunter, *Papermaking*, p. 283.
25. W.H. Smith, *How to Detect Forged Bank Notes*, 1855, p. 11.
26. Patent 12,471, 1849.
27. ibid.
28. Smith, *Forged Bank Notes*, p. 7.
29. Mackenzie, *Bank of England*, pp. 97 & 106.
30. Loeber, *Paper Mould*, p. 53, note 11.
31. Hunter, *Papermaking*, p. 295.
32. Loeber, *Paper Mould*, pp. 33–4.
33. Munsell, *Chronology*, p. 202.
34. W. Green Son & Waite Ltd., 'An Expertise that Makes Its Mark', reprinted from *Paper*.
35. R. Herring, *Paper & Paper Making, Ancient and Modern*, 3 edn. 1863, p. 112.
36. Green, 'Expertise', and R.H. Clapperton, *The Papermaking Machine, Its Invention, Evolution and Development*, 1967, p. 228.

CHAPTER IV   A New Industry Emerges, 1500–1800

1. Sir J. Clapham, *A Concise Economic History of Britain from the Earliest Times to A.D. 1750*, 1949, pp. 262–3.
2. C.P. Hill, *British Economic and Social History, 1700–1964*, 3 edn., 1970, p. 45 and P.S. Bagwell, *The Transport Revolution from 1770*, 1974, p. 47.
3. D. Nuttall, *A Brief History of Platen Presses*, 1979, pp. 1–2.
4. T.K. Derry & M.G. Blakeway, *The Making of Pre-Industrial Britain*, 1969, pp. 195 & 282.
5. R.M. Slythe, *The Art of Illustration, 1750–1900*, 1970, p. 11.
6. ibid, p. 79.
7. J. Krill, *English Artists Paper, Renaissance to Regency*, 1987, pp. 111 & 120.
8. A.H. Shorter, *Paper Making in the British Isles, An Historical and Geographical Study*, 1971, p. 87.
9. ibid, p. 64.
10. Patent 147, 16 Feb. 1665.
11. Patent 284, 6 Nov. 1691.
12. D.C. Coleman, *The British Paper Industry, 1495–1860, A Study in Industrial Growth*, 1958, p. 11.
13. ibid, pp. 15 & 105.
14. ibid, p. 14.
15. Clapham, *Concise History*, p. 186.
16. T.S. Reynolds, *Stronger than a Hundred Men, A History of the Vertical Waterwheel*, 1983, p. 52 and L. Syson, *The Watermills of Britain*, 1980, p. 22.
17. Coleman, *British Paper*, p. 81.
18. E. Hasted, *The History and Topographical Survey of the County of Kent*, Vol II, 1782, p. 132.
19. B. Cooper, *Transformation of a Valley, The Derbyshire Derwent*, 1983, pp. 72 & 132.
20. R. Weible, Ed., *The World of the Industrial Revolution, Comparative and International Aspects of the Industrial Revolution*, 1986, see R.L. Hills, 'Steam and Waterpower – Differences in Transatlantic Approach', pp. 35–53.
21. A.H. Shorter, *Paper Mills and Paper Makers in England, 1495–1800*, 1957, p. 27.
22. T. Churchyard, *A Sparke of Friendship and Warme Goodwill*, 1588, reprinted 1978.
23. Coleman, *British Paper*, p. 51.
24. J. Shaw, *Water Power in Scotland, 1550–1870*, 1984, p. 54 and A.G. Thomson, *The Paper Industry in Scotland, 1590–1861*, 1974, p. 8.
25. Thomson, *Scotland*, p. 10.
26. Shaw, *Water Power*, p. 54.
27. ibid, pp. 55–6.

28. Thomson, *Scotland*, p. 11.
29. Shorter, *Paper Making*, pp. 19 & 22.
30. ibid, p. 75.
31. Coleman, *British Paper*, p. 147.
32. W.A. Churchill, *Watermarks in Paper in Holland, England, France, etc., in the XVII and XVIII Centuries and Their Interconnections*, 1935, pp. 5f.
33. Patent 178, 21 January 1675 and Patent 246, 4 July 1685.
34. Coleman, *British Paper*, pp. 69f. and Shorter, *Paper Making*, p. 25.
35. Thomson, *Scotland*, p. 17.
36. Coleman, *British Paper*, pp. 58–9 and D. Lyddon & P. Marshall, *Paper in Bolton, A Papermaker's Tale*, 1975, p. 23.
37. ibid, p. 98.
38. ibid, pp. 87–8.
39. Shorter, *Paper Making*, p. 67.
40. Shaw, *Water Power*, p. 361.
41. Coleman, *British Paper*, p. 107.
42. D. Hunter, *Papermaking, The History and Technique of an Ancient Craft*, 1978 edn., pp. 312–4.
43. Shorter, *Paper Making*, p. 42.
44. Hunter, *Papermaking*, p. 162.
45. A. Rees, *The Cyclopaedia; or Universal Dictionary of Arts, Sciences and Literature*, 1819, reprint 1972, Vol 4, p. 78 and Thomson, *Scotland*, p. 159.
46. F.M. Bolam, *Stuff Preparation for Paper and Paperboard Making*, 1965, pp. 2–3.
47. Patent 220, 10 July 1682.
48. Patent 242, 11 Oct. 1684.
49. Shorter, *Paper Making*, p. 40.
50. *Encyclopaedia Britannica*, Vol XIII, 1797, p. 709.
51. Patent 1960, 6 Sept. 1793.
52. A.G. Thomson, 'Some Scottish Influences on American Papermaking', *I.P.H. Yearbook*, Vol 4, 1983, p. 299.
53. T. Balston, *James Whatman, Father and Son*, 1957, p. 23 and E.G. Loeber, *Paper Mould and Mouldmaker*, 1982, p. 38.
54. Patent 2045, 1795.
55. Hunter, *Papermaking*, p. 201.
56. Patent 2840, 25 July 1805.
57. Thomson, *Scotland*, p. 49.
58. T. Balston, *William Balston, Paper Maker, 1759–1849*, 1954, pp. 41 & 47.
59. Krill, *Artists Papers*, pp. 33 & 89.
60. Patent 1783, 1790.
61. Coleman, *British Paper*, pp. 122–9.

62. Shorter, *Paper Making*, p. 44.
63. Thomson, *Scotland*, pp. 61 & 66.
64. Coleman, *British Paper*, p. 139.
65. T. Balston, *James Whatman, Father and Son*, 1957, p. 68.
66. Coleman, *British Paper*, p. 136.
67. Balston, *Whatman*, p. 99.
68. Shorter, *Paper Making*, p. 45.

CHAPTER V   The Whatmans and Wove Paper

1. I am indebted to Mr. J. Balston for this information, see also D.C. Coleman, *The British Paper Industry, 1495–1860*, 1958, p. 153.
2. J. Harris, *The History of Kent*, 1719, p. 191.
3. T. Balston, *James Whatman, Father and Son*, 1957, pp. 10–1.
4. ibid, p. 120 and Coleman, *British Paper*, p. 155.
5. Balston, *Whatman*, p. 101.
6. R. Campbell, *The London Tradesman*, 1747, p. 126.
7. K.T. Weiss, *Handbuch der Wasserzeichenkunde*, 1962, p. 171.
8. Balston, *Whatman*, p. 41.
9. Patent No. 1774, 1790.
10. Balston, *Whatman*, pp. 13–4 and E. Hasted, *The History and Topographical Survey of the County of Kent*, Vol II, 1782, p. 132.
11. W. Weiss, 'Concerning Shadowless Laid Handmade Paper', (Typescript), p. 2.
12. Balston, *Whatman*, p. 125.
13. ibid, p. 15.
14. ibid, p. 14.
15. ibid, p. 14.
16. Coleman, *British Paper*, p. 99.
17. J. Krill, *English Artists Paper, Renaissance to Regency*, 1987, p. 74.
18. ibid, p. 75.
19. E. Heawood, *Watermarks, Mainly of the Seventeenth and Eighteenth Centuries*, 1950, see Watermark 3371.
20. Balston, *Whatman*, p. 26.
21. ibid, p. 27.
22. ibid, p. 29.
23. ibid, p. 34.
24. ibid, pp. 152–5.
25. V. Valerio, *L'Italia Nei Manoscritti Dell'Officina Topografica*, 1985, p. 38.
26. Balston, *Whatman*, p. 109 and R.M. Slythe, *The Art of Illustration, 1750–1900*, 1970, p. 86.
27. ibid, p. 110.

28. S. Ireland, *Picturesque Views on the River Medway, with Observations on the Works of Art in its Vicinity*, 1793, p. 118.

29. R. Shepherd, *The Ground and Credibility of the Christian Religion in a Course of Sermons preached before the University of Oxford*, 1788. I am indebted to the Warden of St. Deiniol's Library, Hawarden, for permitting me to look at pre-1800 books in the collections there.

30. M. Koops, *Historical Account of the Substances which have been used to Describe Events and Convey Ideas from the Earliest Date to the Introduction of Paper*, First edn. 1800, 2 edn. 1802.

31. See Balston, *Whatman*, pp. 12–3. For examples of later Whatman and Balston watermarks, see E.W. Brayley, *Thanet and the Cinque Ports*, 1817. The Hollingworth watermark appears as 'Turkey Mills/J. Whatman/1817' in Society of Antiquities, *Sumptibus*, Vol V, plate XXIV, 23 April 1818, and as 'J. Whatman/Turkey Mills/1827 in P. Amsinck, *Tunbridge Wells and its Neighbourhood*, 1810.

32. T. Balston, *William Balston, Paper Maker, 1759–1848*, 1954, pp. 75 & 150 and D. Hunter, *Papermaking, The History and Technique of an Ancient Craft*, 1978 edn., p. 490.

33. J. Whitaker, *The Origin of Arianism Disclosed*, 1791, also in the St. Deiniol's collection.

34. M.J. Fuller, *The Watermills of the East Malling and Wateringbury Streams*, 1980, p. 24 and A.H. Shorter, *Paper Mills and Paper Makers in England, 1495–1800*, 1957, p. 196.

35. ibid, p. 196.

36. ibid, p. 192 and watermark in Society of Antiquaries, *Sumptibus*, Vol IV, plate VII, 8 July 1802, 'Edmeads & Pine 1802'.

37. Shorter, *Paper Mills*, p. 191 and Valerio, *L'Italia*, p. 56.

38. Shorter, *Paper Mills*, p. 186 and Valerio, *L'Italia*, p. 56.

39. Heawood, *Watermarks*, see Watermark 3371.

40. Balston, *Whatman*, p. 35.

41. Shorter, *Paper Mills*, p. 137 and for watermark see *Antiquarian and Topographical Researches of the Several Most Interesting Objects of Curiosity in Great Britain*, 1817.

42. Shorter, *Paper Mills*, p. 124, and for watermark see Brayley, *Thanet*.

43. J. Baxter, Ed., *The Sister Arts, or a Concise and Interesting View of the Nature and History of Paper-Making, Printing and Bookbinding*, 1809, p. 21.

44. A. Rees, *The Cyclopaedia, or Universal Dictionary of Arts, Sciences and Literature*, 1819, article on 'Paper'.

45. Hunter, *Papermaking*, p. 129.

46. E.G. Loeber, *Paper Mould and Mouldmaker*, 1982, p. 23.

47. Coleman, *British paper*, p. 90.

48. Koops, *Historical Account*, 1800, p. 72.

CHAPTER VI   The Origins of Wallpaper

1. A.V. Sugden & J.L. Edmondson, *A History of English Wallpaper, 1509–1914*, c. 1925, p. 2.
2. ibid, p. 5.
3. C.C. Oman & J. Hamilton, *Wallpapers, A History and Illustrated Catalogue of the Collection in the Victoria and Albert Museum*, 1982, p. 9.
4. Sugden, *History*, p. 9 and F. Teynac, P. Nolot and J.-D. Vivien, *Wallpaper, a History*, 1982, p. 18.
5. Oman, *Wallpapers*, p. 9 and Anon, *A Decorative Art, 19th Century Wallpapers in The Whitworth Art Gallery*, 1985, p. 19 (This will be referred to as 'Whitworth' subsequently).
6. Sugden, *History*, p. 7.
7. ibid, p. 9.
8. Oman, *Wallpapers*, p. 16.
9. ibid, p. 14.
10. Sugden, *History*, pp. 85 & 102.
11. Oman, *Wallpapers*, p. 32.
12. Sugden, *History*, p. 97.
13. ibid, p. 106.
14. *Erddig House, Clwyd*, 1978, A Guide Book published by the National Trust. I am most grateful to the Administrator for showing me these papers.
15. Communication from the paper conservator, Mrs. MacDermott-Topping, Newcastle-upon-Tyne.
16. Oman, *Wallpapers*, p. 34.
17. Whitworth, *Decorative Art*, p. 5.
18. Sugden, *History*, p. 56.
19. ibid, p. 46 and Teynac, *History*, p. 45.
20. Sir J. Clapham, *A Concise Economic History of Britain From the Earliest Times to A.D. 1750*, 1949, p. 186.
21. Sugden, *History*, p. 47.
22. ibid, p. 138.
23. Oman, *Wallpapers*, p. 39.
24. Sugden, *History*, p. 139.
25. ibid, p. 56.
26. Oman, *Wallpapers*, p. 35.
27. J. Munsell, *Chronology of the Origin and Progress of Paper and Paper-Making*, 5 edn, 1874, reprint 1980, p. 49. and Sugden, *History*, p. 119.
28. Sugden, *History*, p. 120 and Teynac, *History*, p. 98.
29. Munsell, *Chronology*, p. 139.
30. Whitworth, *Decorative Art*, pp. 5–6.

31. A. Rees, *The Cyclopaedia; or Universal Dictionary of Arts, Sciences and Literature*, 1819, reprint 1972, Vol IV, p. 89.
32. This information is derived from studying the collections of wallpapers at both the Whitworth Art Gallery, Manchester, and the Victoria and Albert Museum, London. I am most grateful to the Keepers at these museums for their assistance.
33. Whitworth, *Decorative Art*, p. 18.
34. Sugden, *History*, p. 124.
35. ibid, p. 114.
36. Whitworth, *Decorative Art*, p. 10.
37. Teynac, *History*, p. 129.
38. *Journal of Design*, Vol. IV, 1850–1, p. 174.
39. C. Tomlinson, *Cyclopaedia of Useful Arts & Manufactures*, 1854, p. 376.
40. Whitworth, *Decorative Art*, p. 11.

CHAPTER VII   The Development of the Fourdrinier Paper Machine

1. R.H. Clapperton, *The Paper-making Machine, Its Invention, Evolution and Development*, 1967, p. 16. This gives the fullest account of the invention and development of the early papermachines which is available anywhere.
2. D. Hunter, *Papermaking, The History and Technique of an Ancient Craft*, 1978 edn., p. 348.
3. A.H. Shorter, *Paper Making in the British Isles, An Historical and Geographical Study*, 1971, p. 98.
4. Clapperton, *Machine*, p. 30.
5. The following are the appropriate English patents; No. 2487, 20 April 1801 by John Gamble, No. 2708, 5 December 1803 by John Gamble and No. 3068, 14 August 1807 by Henry Fourdrinier, Sealy Fourdrinier and John Gamble.
6. S.B. Donkin, 'Bryan Donkin, F.R.S., M.I.C.E., 1768–1855', *Trans. Newcomen Soc.*, Vol XXVII, 1949–50 & 50–51, p. 86.
7. Clapperton, *Machine*, p. 41.
8. The Bryan Donkin Co. Ltd., *A Brief Account of Bryan Donkin, F.R.S., and of the Company he Founded 150 Years Ago*, p. 12.
9. Clapperton, *Machine*, pp. 31 & 33.
10. ibid, p. 47.
11. D.C. Coleman, *The British Paper Industry, 1495–1860*, 1958, p. 186.
12. A.G. Thomson, *The Paper Industry in Scotland, 1590–1861*, 1974, p. 177.
13. Donkin, 'Bryan Donkin', p. 87.
14. T. Balston, *William Balston, Paper Maker, 1759–1849*, 1954, p. 52.
15. J. Munsell, *Chronology of the Origin and Progress of Paper and Paper-Making*, 5th edn. 1876, reprinted 1980, pp. 61–2.
16. Coleman, *Paper Industry*, p. 198.
17. Thomson, *Scotland*, p. 169.

18. ibid, p. 190.

CHAPTER VIII   Rival Machines for Making Paper

1. H. Voorn, 'Machines for Producing Paper in Sheets', *I.P.H. Information*, 1987, No 3, p. 124.
2. Patent 2840, 25 July 1805.
3. R.H. Clapperton, *The Papermaking Machine, Its Invention, Evolution and Development*, 1967, pp. 11 & 271.
4. J. van Houtum, 'The Development of the Cylinder Mould Machine for Making True Watermarks', *I.P.H. Year Book*, Vol 5, 1984, p. 155.
5. I. McNeil, *Joseph Bramah, A Century of Invention, 1749–1851*, 1968, p. 141.
6. Clapperton, *Machine*, p. 54 and Patent 2951, July 1806.
7. Patent 3084, December 1807 and No. 3580, July 1812.
8. Clapperton, *Machine*, pp. 58–61.
9. ibid, p. 56.
10. Patent 4002, March 1816.
11. A.G. Thomson, *The Paper Industry in Scotland, 1590–1861*, 1974, pp. 161–3.
12. van Houtum, 'Cylinder Mould Machine', p. 157.
13. E.G. Loeber, *Paper Mould and Mould Maker*, 1982, p. 44.
14. Voorn, 'Machines', p. 130.
15. Clapperton, *Machine*, p. 331.
16. Patent 3080, 12 Nov. 1807.
17. Patent 3056, 30 June 1807.
18. Patent 3191, 19 January 1809.
19. A.H. Shorter, *Paper Making in the British Isles, An Historical and Geographical Study*, 1971, p. 119.
20. Patent 3191, 1809, pp. 3–4 of 1856 printed version.
21. van Houtum, 'Cylinder Mould Machine', p. 156.
22. Patent 3191, p. 9.
23. Patent 3425, 21 May 1811.
24. P. Barlow, *A Treatise on the Manufactures and Machinery of Great Britain*, 1836, p. 769.
25. J. Evans, *The Endless Web, John Dickinson & Co. Ltd., 1804–1954*, 1955, p. 11.
26. ibid, p. 21, quoted from The Rev. T.F. Dibdin, *Bibliographical Decameron*, 1817.
27. ibid, p. 38.
28. Clapperton, *Machine*, p. 83.
29. Patent 4509, 1820, enrolled 28 April 1821.
30. D. Lyddon & P. Marshall, *Paper In Bolton, A Papermaker's Tale*, 1975, p. 37.
31. L.N. Burt, 'Watford and Papermaking', *Trans. Newcomen Soc.*, Vol XXV, 1945–6 & 46–7, p. 106.
32. Evans, *Endless Web*, p. 11.

33. D. Hunter, *Papermaking, The History and Technique of an Ancient Craft*, 1978 edn., p. 351.
34. Thomson, *Scotland*, p. 158.
35. Clapperton, *Machine*, p. 345.
36. Hunter, *Papermaking*, pp. 353–4.
37. A.G. Thomson, 'Some Scottish Influences on American Papermaking', *I.P.H. Year Book*, Vol 4, 1983, p. 302.
38. Hunter, *Papermaking*, p. 355.
39. J. Munsell, *Chronology of the Origin and Progress of Paper and Paper-Making*, 5th edn. 1876, reprint 1980, pp. 85–6.
40. Voorn, 'Machines', p. 128.
41. Patent 5754, 14 July 1829.

CHAPTER IX   The Nineteenth Century; Demand Outstrips Rag Supplies

1. D.C. Coleman, *The British Paper Industry, 1495–1860, A Study in Industrial Growth*, 1958, p. 209 and T.K. Derry & T.I. Williams, *A Short History of Technology from Earliest Times to A.D. 1900*, 1960, p. 644.
2. D. Nuttall, *A Brief History of Platen Presses*, 1979, pp. 4–7.
3. J. Moran, *Printing Presses, History and Development from the Fifteenth Century to Modern Times*, 1973, pp. 135 and 137.
4. Derry & Williams, *Short History*, p. 646.
5. Coleman, *British Paper*, pp. 209–10.
6. P. Barlow, *A Treatise on the Manufactures and Machinery of Great Britain*, 1836, p. 779.
7. Derry & Williams, *Short History*, p. 647.
8. C. Tomlinson, Ed., *Cyclopaedia of Useful Arts and Manufactures*, 1854, pp. 490–1.
9. ibid, p. 596.
10. R. Herring, *Paper and Paper Making, Ancient and Modern*, 3 edn. 1863, p. 31.
11. Moran, *Printing Presses*, p. 212.
12. Derry & Williams, *Short History*, p. 641.
13. R.H. Clapperton, *The Paper-Making Machine, its Invention, Evolution and Development*, 1967, p. 292.
14. Tomlinson, *Cyclopaedia*, p. 490.
15. J. Evans, *The Endless Web, John Dickinson & Co. Ltd., 1804–1954*, 1955, pp. 72–3.
16. ibid, pp. 81–3.
17. ibid, p. 72.
18. Patent 10,565, March 1845.
19. Evans, *Endless Web*, p. 135.
20. ibid, p. 80.
21. Coleman, *British Paper*, p. 317.
22. ibid, p. 318.

23. ibid, p. 323.
24. A.H. Shorter, *Paper Making in the British Isles, An Historical and Geographical Study*, 1971, p. 117.
25. Coleman, *British Paper*, p. 202.
26. Barlow, *Treatise*, p. 769, Herring, *Paper*, p. 125 and J. Munsell, *Chronology of the Origin and Progress of Paper and Paper-Making*, 5 edn. 1876, reprint 1980, p. various.
27. J. Shaw, *Water Power in Scotland, 1550–1870*, 1984, pp. 361–2.
28. Munsell, *Chronology*, p. 59.
29. Herring, *Paper*, p. 71.
30. Coleman, *British Paper*, p. 214.
31. ibid, p. 201.
32. Evans, *Endless Web*, p. 107.
33. Shorter, *Paper Making*, p. 139.
34. Coleman, *British Paper*, p. 215.
35. Evans, *Endless Web*, p. 86.
36. A. Rees, *The Cyclopaedia; or Universal Dictionary of Arts, Science and Literature*, 1819, reprint 1972, Vol IV, p. 74.
37. Munsell, *Chronology*, p. 106.
38. Coleman, *British Paper*, p. 338.
39. J. Krill, *English Artists Papers, Renaissance to Regency*, 1987, p. 141.
40. A.G. Thomson, *The Paper Industry in Scotland, 1590–1870*, 1974, p. 36.
41. A.G. Thomson, 'Some Scottish Influences on American Papermaking', *I.P.H. Year Book*, Vol 4, 1983, p. 300.
42. M. Tillmanns, *Bridge Hall Mills, Three Centuries of Paper and Cellulose Film Manufacture*, 1978, p. 16.
43. Patent 1872, 25 April 1792. See also Patent 1922, 28 November 1792, of Hector Campbell for a similar method by boiling rags first in a caustic ley and then using bleaching powders.
44. T. Balston, *James Whatman, Father and Son*, 1957, p. 103.
45. Coleman, *British Paper*, p. 216 and T.I. Williams, *A Biographical Dictionary of Scientists*, 1969, p. 506.
46. Krill, *Artists Papers*, p. 142.
47. Munsell, *Chronology*, pp. 70 & 98.
48. Thomson, *Scotland*, p. 38.
49. Munsell, *Chronology*, p. 68 and Rees, *Cyclopaedia*, p. 75.
50. M. Koops, *Historical Account of the Substances which have been used to Describe Events, and to Convey Ideas from the earliest date to the Invention of Paper*, 2 edn. 1801, p. 250.
51. Patent 2392, 28 April 1800.
52. Patent 2433, 2 August 1800.
53. Patent 2481, 17 February 1801.

54. Koops, *Historical Account*, 1 edn., 1800, p. 10.
55. D. Hunter, *Papermaking, the History and Technique of an Ancient Craft*, 1978 edn., pp. 201 & 335.
56. ibid, pp. 336 & 339.
57. ibid, p. 338.
58. Patent 5041, 1824.
59. Patent 964, 28 April 1854.
60. Patent 1039, 1856.
61. F. Bolam, Ed., *Paper Making, A General Account of its History, Processes and Applications*, 1965 edn., p. 127.
62. Evans, *Endless Web*, p. 110.
63. F.H. Norris, *Paper and Paper Making*, 1952, p. 28.
64. Herring, *Paper*, pp. 67–8.
65. ibid, pp. 69–70.

CHAPTER X   Esparto and Other Types of Pulp

1. F. Bolam, Ed., *Paper Making, A General Account of Its History, Processes and Applications*, 1965 edn., p. 36.
2. Patent 8273, 19 Nov. 1839.
3. J. Evans, *The Endless Web, John Dickinson & Co. Ltd., 1804–1954*, 1955, p. 111.
4. Patent 1452, 1853 and Patent 294, 7 February 1854.
5. Patent 1816, July 1856 and Patent 274, February 1860.
6. D.C. Coleman, *The British Paper Industry, 1495–1860, A Study in Industrial Growth*, 1958, p. 216 and A.H. Shorter, *Paper Making in the British Isles, An Historical and Geographical Study*, 1971, p. 139.
7. Coleman, *British Paper*, p. 342 and M. Tillmanns, *Bridge Hall Mills, Three Centuries of Paper and Cellulose Film Manufacture*, 1978, p. 50.
8. Evans, *Endless Web*, pp. 111 & 112.
9. ibid, p. 111.
10. J. Munsell, *Chronology of the Origin and Progress of Paper and Paper-Making*, 5 edn. 1876, reprint 1980, p. 183.
11. M. Chater, *Family Business, A History of Grosvenor Chater, 1690–1977*, 1977, pp. 21 & 26.
12. A.E. Davies, 'Paper-Mills and Paper-Makers in Wales, 1700–1900', *The National Library of Wales Journal*, Vol XV, 1967–8, p. 15.
13. A.G. Thomson, *The Paper Industry in Scotland, 1590–1861*, 1974, p. 201.
14. J. Shaw, *Water Power in Scotland, 1550–1870*, 1984, p. 513.
15. Shorter, *Paper Making*, p. 218.
16. C.F. Cross & E.J. Bevan, *A Text-Book of Paper-Making*, 1916, p. 472.
17. Cross & Bevan, *Text-Book*, p. 130 and Bolam, *Paper Making*, p. 53.

18. Patent 1816, 1856.
19. Patent 532, 25 February 1865, Patent 1602, 13 June 1865 and Patent 812, 12 March 1866.
20. R.W. Sindall, *Paper Technology, An Elementary Manual on the Manufacture, Physical Qualities and Chemical Constituents of Paper and of Paper-Making Fibres*, 2 edn., 1910, p. 37.
21. Shorter, *Paper Making*, p. 165.
22. M. Wray, *The British Paper Industry, A Study in Structural and Technological Change*, 1979, p. 116.
23. Munsell, *Chronology*, p. 78.
24. Patent 7643, 13 November 1838.
25. D. Hunter, *Papermaking, The History and Technique of an Ancient Craft*, 1978 edn., pp. 376 & 380.
26. Cross & Bevan, *Text-Book*, p. 166.
27. ibid, p. 167.
28. Sindall, *Paper Technology*, p. 54.
29. Cross & Bevan, *Text-Book*, p. 167.
30. J. Grant & J.H. Young, *Paper and Board Manufacture, A General Account of Its History, Processes and Applications*, 1978, p. 75.
31. Wray, *British Paper*, p. 113.
32. Hunter, *Papermaking*, p. 390.
33. Cross & Bevan, *Text-Book*, p. 147.
34. F.H. Norris, *Paper and Paper Making*, 1952, p. 54 and Sindall, *Paper Technology*, p. 67.
35. R.H. Clapperton & W. Henderson, *Modern Paper-Making*, 1929, p. 62.
36. Hunter, *Papermaking*, p. 575.
37. Sindall, *Paper Technology*, p. 67.
38. Hunter, *Papermaking*, pp. 391–2.
39. Shorter, *Paper Making*, p. 141.
40. Cross & Bevan, *Text-Book*, p. 152.
41. Sindall, *Paper Technology*, p. 63.
42. Bolam, *Paper Making*, p. 95.
43. Grant & Young, *Paper and Board*, p. 83.
44. Cross & Bevan, *Text-Book*, pp. 473–5.
45. Wray, *British Paper*, p. 111 and British Paper and Board Industry Federation, *Reference Tables*, 1986, p. 78.

CHAPTER XI   The Paper Machine Matures

1. R.H. Clapperton, *The Paper-making Machine, Its Invention, Evolution and Development*, 1967, p. 215 and A.H. Shorter, *Paper Making in the British Isles, An Historical and Geographical Study*, 1971, pp. 124–7.

2. J. Evans, *The Endless Web, John Dickinson & Co. Ltd., 1804–1954*, 1955, p. 13.
3. D.C. Coleman, *The British Paper Industry, 1495–1860, A Study in Industrial Growth*, 1958, p. 198.
4. J. Munsell, *Chronology of the Origin and Progress of Paper and Paper-Making*, 5 edn. 1876, reprint 1980, pp. 108 & 119.
5. Shorter, *Paper Making*, p. 101.
6. D.K. Clark, *The Exhibited Machinery of 1862*, 1864, p. 231 and D. Lyddon & P. Marshall, *Paper in Bolton, A Papermaker's Tale*, 1975, pp. 93 & 113.
7. I wish to thank Messrs. J. & J. Makin for making their archives available which supplied this information.
8. D. Hunter, *Papermaking, The History and Technique of an Ancient Craft*, 1978 edn., p. 577.
9. Clark, *Machinery*, pp. 231–3 and R.W. Sindall, *Paper Technology, An Elementary Manual on the Manufacture, Physical Qualities and Chemical Constituents of Paper and of Paper-Making Fibres*, 1910, p. 72.
10. A.G. Thomson, *The Paper Industry in Scotland, 1590–1861*, 1974, p. 182 and Evans, *Endless Web*, p. 13.
11. Shorter, *Paper Making*, p. 109.
12. M. Tillmanns, *Bridge Hall Mills, Three Centuries of Paper and Cellulose Film Manufacture*, 1978, p. 25.
13. Clark, *Machinery*, p. 236 and Coleman, *British Paper*, p. 213.
14. ibid, p. 227.
15. Evans, *Endless Web*, p. 45.
16. J. Shaw, *Water Power in Scotland, 1550–1870*, 1984, p. 512.
17. Tillmanns, *Bridge Hall Mills*, p. 81.
18. Evans, *Endless Web*, p. 129.
19. Lyddon & Marshall, *Paper*, p. 115.
20. Shorter, *Paper Making*, p. 38.
21. ibid, p. 110.
22. Thomson, *Scotland*, pp. 55 & 159.
23. Coleman, *British Paper*, p. 231.
24. T. Balston, *William Balston, Paper Maker, 1759–1849*, 1954, p. 41.
25. Tillmanns, *Bridge Hall Mills*, p. 27.
26. Evans, *Endless Web*, p. 96.
27. J. & J. Makin Records, 'Valuation of the North Wales Paper Company's Property at Oakenholt, near Flint'.
28. Evans, *Endless Web*, p. 174.
29. J. & J. Makin Records, 'Reed's Patent Beater', 28 March 1892.
30. Patent 8283, 1839, enrolled 25 May 1840.
31. Patent 2828, 1856.
32. Patent 792, 27 March 1860 and Patent 2019, 22 August 1860.
33. Patent 4152, 1817, enrolled 5 February 1818.

34. Patent 6866, 1835, enrolled 21 February 1836.
35. Patent 8212, enrolled 1 November 1839.
36. Patent 5964, 1830.
37. Patent 6148, 1831.
38. Patent 6209, 1832.
39. Clapperton, *Machine*, p. 316 f. for one developed by Bryan Donkin and G.E. Sellers from America.
40. Patent 5617, enrolled 20 June 1828.
41. Patent 108, 1858 and Clapperton, *Machine*, p. 193.
42. Patent 5617, enrolled 20 June 1828.
43. Clapperton, *Machine*, pp. 101–2 & 250–2.
44. Patent 7098, enrolled 14 November 1836.
45. J. & J. Makin Records, 'Paper Machine Specification', 15 December 1886.
46. Clapperton, *Machine*, pp. 150–3.
47. Lyddon & Marshall, *Paper*, pp. 90–1 and Patent 11,833, 1847, enrolled 28 January 1848.
48. M. Bailey, 'Robert Stephenson & Company, 1823–1836', M.Sc. Thesis, Newcastle upon Tyne 1984, pp. 69–70 & 315.
49. G.T. Mandl, *Three Hundred Years in Paper*, 1985, pp. 92 f. and Patent 13,171, 1850.
50. Patent 4152, enrolled 5 February 1818.
51. Patent 6008, 6 April 1831.
52. Clapperton, *Machine*, pp. 214 & 244.
53. J. Krill, *English Artists Paper, Renaissance to Regency*, 1987, p. 30.
54. Thomson, *Scotland*, p. 50.
55. R. Herring, *Paper & Paper Making, Ancient and Modern*, 3 edn., 1863, p. 92.
56. Clapperton, *Machine*, p. 115.
57. ibid, p. 154.
58. Patent 13,171, 1850.
59. Tillmanns, *Bridge Hall Mills*, p. 61.
60. M.J. Fuller, *The Water Mills of the East Malling and Wateringbury Streams*, 1980, p. 55.
61. J. Murray, *Practical Remarks on Modern Paper*, 1829, pp. 66–7.
62. Thomson, *Scotland*, p. 189.
63. Clapperton, *Machine*, p. 95 and Thomson, *Scotland*, p. 164.
64. Hunter, *Papermaking*, p. 555.
65. Bailey, 'Stephenson', p. 21.
66. Clapperton, *Machine*, pp. 97 f. and A. Muir, *The Kenyon Tradition, The History of James Kenyon & Son Ltd., 1664–1964*, 1964, pp. 36 & 39.
67. Evans, *Endless Web*, p. 22 and Thomson, *Scotland*, p. 166.

CHAPTER XII   Watermarks on Papermachines

1. Patent 3068, 1807.
2. Patent 5380, 1826.
3. R.H. Clapperton, *The Paper-making Machine, Its Invention, Evolution and Development*, 1967, pp. 109–12.
4. Patent 3191, 1809.
5. J. Evans, *The Endless Web, John Dickinson & Co. Ltd., 1804–1954*, 1955, pp. 42–3.
6. A.D. Mackenzie, *The Bank of England Note, A History of Its Printing*, 1953, p. 146.
7. J. Van Houtum, 'The Development of the Cylinder Mould Machine for Making True Watermarks', *I.P.H. Year Book*, Vol 5, 1984, p. 159.
8. Clapperton, *Machine*, p. 96.
9. Patent 5075, 1825.
10. D. Hunter, *Papermaking, The History and Technique of an Ancient Craft*, 1978 edn., pp. 400–1.
11. ibid, p. 95.
12. Patent 5647, 1828, and see Clapperton, *Machine*, p. 99.
13. Clapperton, *Machine*, Plate No. 26.
14. Patent 5934, 1830, and see Clapperton, *Machine*, p. 226.
15. Patent 7977, 1839.
16. ibid.
17. I am indebted to Messrs. W. Green Son & Waite Ltd. for their help in understanding the latest developments in the manufacture of dandy rolls.
18. W. Green Son & Waite Ltd., 'An Expertise that Makes its Mark', reprinted from *Paper*.

CHAPTER XIII   The Twentieth Century

1. M. Wray, *The British Paper Industry; A Study in Structural and Technological Change*, 1979, various tables.
2. British Paper and Board Industry Federation, *Reference Tables, 1986*, p. 113.
3. B.P. & B.I.F., *Tables*, p. 2 and Wray, *British Paper*, p. 219.
4. Wray, *British Paper*, p. 34.
5. A. Muir, *The British Paper and Board Maker's Association, 1872–1972*, 1972, p. 53.
6. Wray, *British Paper*, p. 91.
7. B.P. & B.I.F., *Tables*, p. 38.
8. M. Tillmanns, *Bridge Hall Mills, Three Centuries of Paper & Cellulose Film Manufacture*, 1978, p. 93.
9. I wish to thank Mr. Alan Marriott of the B.P. & B.I.F. for this information.
10. *Daily Telegraph*, 12 August, 1987.
11. J. Green, *Yates Duxbury & Sons, Papermakers of Bury*, 1963, p. 31.

12. J. Evans, *The Endless Web, John Dickinson & Co. Ltd., 1804–1954*, 1955, pp. 111 & 208.
13. C.P. Klass, *Cylinder Board Manufacture*, 1966, p. 15.
14. ibid, p. 15.
15. W.J. Carter, Ed., *Papermachine Clothing*, 1974, p. 17.
16. T.I. Williams, Ed., *A History of Technology, Vol VI, The Twentieth Century, c. 1900 to c. 1950*, Part I, 1978, p. 620.
17. Carter, *Clothing*, p. xi.
18. J. Grant & J.H. Young, Ed., *Paper and Board Manufacture, A General Account of Its History, Processes and Applications*, 1978, p. 224.
19. R.H. Clapperton, *The Paper-making Machine, Its Invention, Evolution and Development*, 1967, pp. 231.
20. Wray, *British Paper*, pp. 103 & 123.
21. F.H. Norris, *Paper and Paper Making*, 1952, p. 145.
22. A. Muir, *The Kenyon Tradition, The History of James Kenyon & Son Ltd., 1664–1964*, 1964, pp. 99 & 101.
23. Carter, *Clothing*, p. 115.
24. R.H. Clapperton & W. Henderson, *Modern Paper-making*, 1929, p. 184.
25. B.W. Attwood, 'Papermaking as a technology: A survey with comments on the British contribution', *Paper Technology*, 1969, Vol 10, No. 5, p. 364.
26. Clapperton, *Machine*, p. 170.
27. Norris, *Paper*, p. 205.
28. D. Lyddon & P. Marshall, *Paper in Bolton, A Papermaker's Tale*, 1975, p. 173.
29. J. & J. Makin Records, 'J. & J. Makin Ltd., 1887–1984'.
30. Wray, *British Paper*, p. 105.
31. See De Juliius Spa Macchine per Cartiere, *Secoli di Tradizione Nei Nostri Trent'Anni di Esperienza*, 1987 and communication from Mr. B.W. Attwood.
32. Attwood, 'Papermaking', p. 364.
33. Wray, *British Paper*, p. 107.
34. *Newsprint from Shotton.*
35. *Paper UK*, 'The Caledonian Papermill', 20 April, 1987, p. 2.
36. Wray, *British Paper*, p. 38.
37. B.P. & B.I.F. *Fact Sheet*, No. 38.
38. Wray, *British Paper*, p. 115 and B.P. & B.I.F. *Tables*, p. 93.
39. Grant & Young, *Paper*, p. 140.
40. B.P. & B.I.F. *Fact Sheet* No. 22.
41. Kimberly-Clark, *UK Tissue Market Report*, 1986, p. 2.

INDEX